Edexcel AS
BIOLOGY
With Stafford

Paper code: 9BN0/01:
Topic 1: Lifestyle, Health and Risk
Topic 2: Genes and Health

And

IAS Unit code: WBI11/01
Unit 1: Molecules, Diet, Transport and Health

ISBN: 978-81-910705-9-0

Typeset & layouts by: Dr. Stafford Valentine Redden
Cover designed by: Pro-designers, Maldives
Printed by: Niswass Press, Bhubaneswar
Printed Internationally by: Createspace.com, USA
Published by: Author publisher
Date : April 2018

ISBN: 978-81-910705-9-0

Contact details of the author:
Stafford Valentine Redden
Near Hindhustan Oil Mill,
Ramchandrapur, Jatni P.O,
Orissa, India, 752050
staffordv@yahoo.com

CONTENTS

CHAPTER ONE

WATER – THE UNIVERSAL SOLVENT

Edexcel Specification 1.1: *Understand the importance of water as a solvent in transport, including its dipole nature.*

Dipolar nature of water

The dipole nature of water means that one end of the water molecule bears a slight negative charge, while the other end bears a slight positive charge, as shown in fig.1.1.

Figure 1.1: Dipole water molecule

Water is a polar covalent compound. Hydrogen and oxygen are sharing electrons. Oxygen has a greater electronegativity than hydrogen does. This means that oxygen pulls the shared pair of electrons closer to itself and thereby bears a slight negative charge while the hydrogen side of the water molecule bears a slight positive charge. Hence water is dipolar.

Formation of hydrogen bonds

Due to the dipolar nature of water, individual water molecules form hydrogen bonds with each other. The positive end of one water molecule is attracted to the negative end of another water molecule. This inter-molecular force of attraction is called a hydrogen bond, shown in figure 1.2.

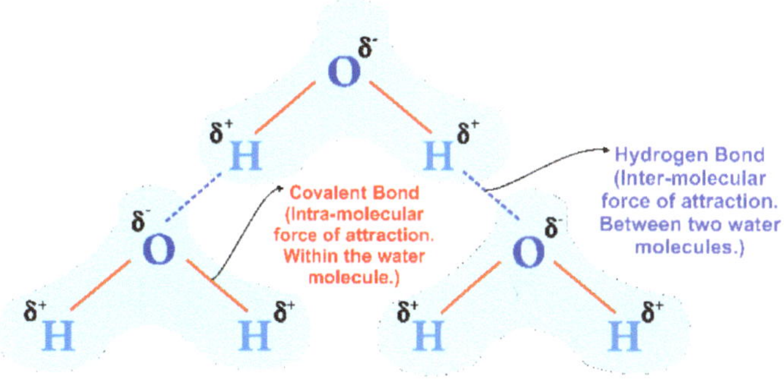

Figure 1.2: Hydrogen bonding between water molecules

The dipole nature and hydrogen bonding are responsible for most of the properties of water that make it useful for life.

Water: The universal solvent

Water is considered the universal solvent because ionic solutes, polar covalent solutes and large complex covalent solutes can dissolve in water.

1. ***Dissolving ionic solutes***

The dipolar nature of water molecule makes it a good solvent. For example, common table salt, sodium chloride, is an ionic substance that contains alternating sodium and chloride ions held together in an ionic lattice.

The partial charge that develops across the water molecule helps to make it an excellent solvent. When sodium chloride is added to water, the partial charges on the water molecules are attracted to the Na^+ ions and Cl^- ions. The water molecules work their way into the crystal structure and between the individual ions, surrounding them by a **hydration shell** and slowly dissolve the salt. The negative oxygen ends of water molecules will surround the positive sodium ions; the positive hydrogen ends of water molecules will surround the negative chloride ions, as shown in figure 1.3.

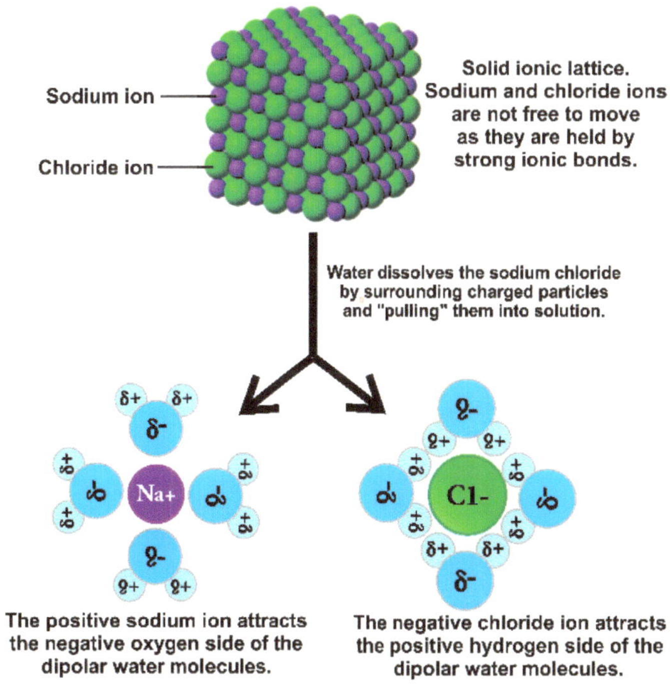

Figure 1.3: Dissolving ionic solutes

2. *Dissolving polar covalent solutes*

Any molecule with polar regions will dissolve in water. Glucose is a large covalent molecule. However, there are many polar groups on its surface. The water molecules will be attracted to these polar groups and pull the glucose molecules away from the surface of the crystal to form a solution, as shown in figure 1.4.

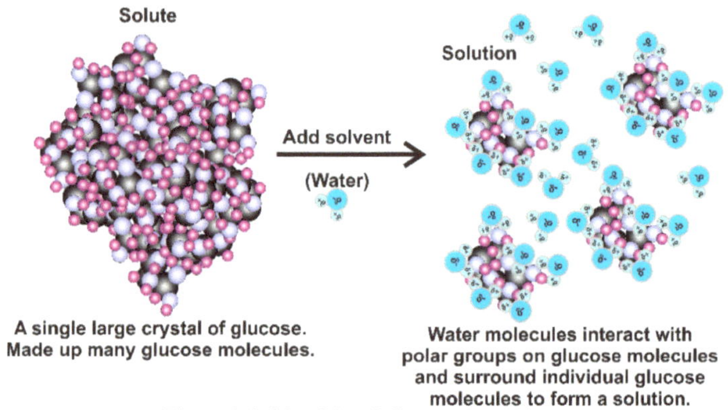

Figure 1.4: Dissolving Polar covalent solutes.

3. *Dissolving large complex solutes, such as proteins*

Any molecule with polar regions or charges on its surface will dissolve in water. Water molecules gather around large protein molecules to form a special kind of solution called a **colloid**, as found in the cytoplasm of cells. The charged regions on the protein are called as hydrophilic regions as they attract water molecules as shown in figure 1.5. These hydrophilic interactions play an important role in the maintenance of the tertiary structure of enzymes.

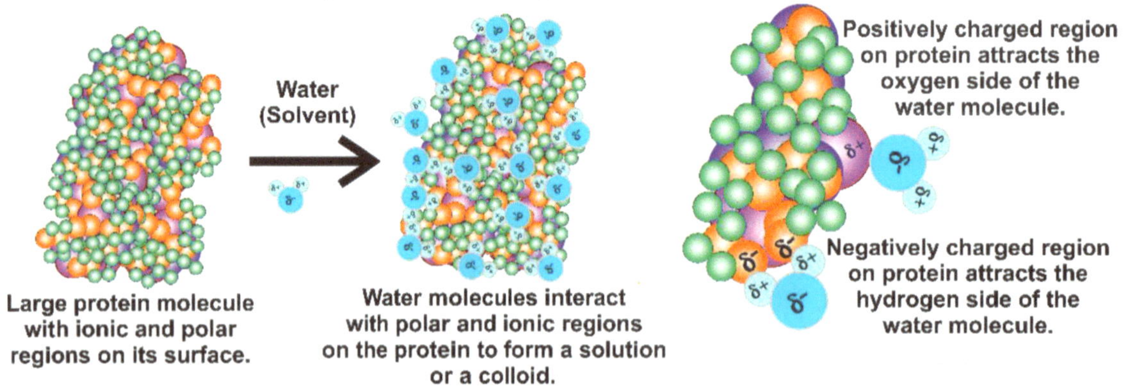

Figure 1.5: Dissolving Proteins in water due to hydrophilic interactions.

Hydrophobic interactions prevent a substance from dissolving in water

Nonpolar and nonionic substances do not dissolve in water. Oil, for example, is a nonpolar molecule. Because there is no net electrical charge across an oil molecule, it is not attracted to water molecules and therefore does not dissolve in water. Hence it is called hydrophobic (Water hating).

Water as a useful solvent in biological systems

1. Transport medium

- **Sap is transported** through xylem and phloem of plants. The sap consists of sucrose, amino acids and ions **dissolved in water,** as shown in figure 1.6.

Figure 1.6: Transport of sap as aqueous solution in plants

- Glucose, amino acids, hormones, urea and many other substances **dissolve in the blood plasma** and can be easily **transported in the blood,** as shown in figure 1.7.

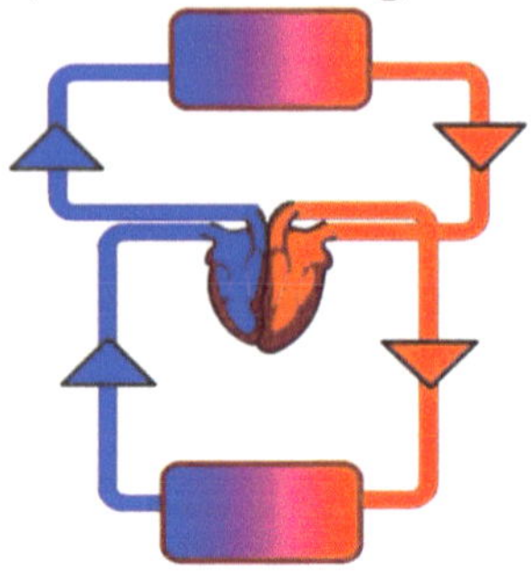

Figure 1.7: Transport of solutes as an aqueous solution in the blood

2. *Medium to facilitate collisions between reactants*

Water also provides a good medium for soluble substances to collide with each other and react. Enzymes are soluble in water and collide with substrates to bring about metabolic reactions. These collisions would not be possible or would be too slow in solid state.

3. *Cohesion and adhesion: a consequence of the dipolar nature of water*

- Cohesion is the force of attraction between molecules of the same substance. The dipolar nature of water makes it very cohesive. This allows water to be pulled along the xylem vessels with relative ease.
- Adhesion is the force of attraction between molecules of different substances. Water is also a good adhesive. It will cling onto many objects and act as a glue. Capillary Action is an example of cohesion and adhesion working together to move water up a thin tube like the xylem in plants.

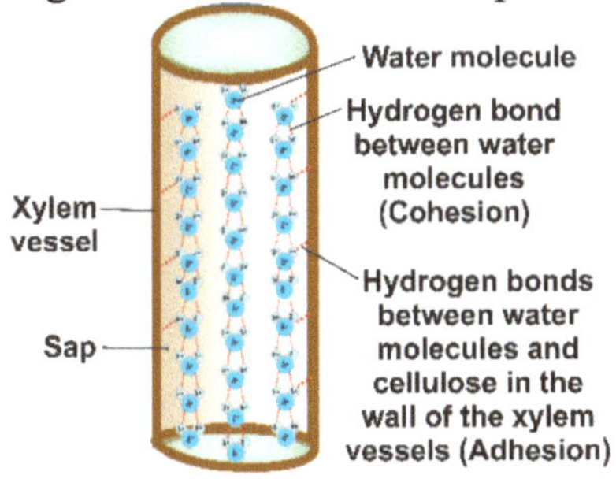

Figure 1.8: Cohesion and adhesion in a xylem vessel

4. ***Water has a high latent heat of vapourisation:*** The latent heat of vapourisation is the amount of heat energy needed to vapourise 1 kilogram of water at a constant temperature. In humans (at 37 ^{0}C), this value is 539 000 cal/Kg. So, when water evaporates from a surface it absorbs latent heat of vapourisation from the surface and cools the surface down. This makes water a good coolant. This is especially useful in cooling of plant tissues by transpiration and cooling of mammals by sweating or panting.

Figure 1.9: Transpiration, sweating and panting help to keep organisms cool

5. ***Water has a high latent heat of Fusion:*** The latent heat of fusion is the amount of heat energy needed to convert 1 kilogram of water into ice at a constant temperature. This value for water is about 80 KJ/g. The high latent heat of fusion ensures that water bodies do not freeze easily in cold climates. It also prevents the cytoplasm from freezing at low temperatures.

6. ***Water has a high specific heat capacity:*** The specific heat capacity is the amount of heat energy needed to increase the temperature of 1 kilogram of water by 1^{0}C. The value for water is 4200 J/Kg ^{0}C. So, a lot of heat must be added or removed to change the temperature of water. This property of water prevents sudden fluctuations in temperature of organisms or aquatic environments, allowing enzymes to function at their optimum temperature. This high specific heat allows water to act as a heat sink. Water will retain its temperature after absorbing large amounts of heat and retain its temperature after losing equally large amounts of heat. The reason for this is that water has many hydrogen bonds. A lot of heat energy is involved in breaking these bonds.

7. *Surface tension* is the property of a liquid which makes its surface behave like a stretched membrane, mainly caused due to hydrogen bonding between water molecules, as shown in figure 1.9. This allows materials to rest upon it if the surface tension is not broken. Pollen, dust, water insects, and other biological material can remain on the surface of the water because of this tension. Mosquito larvae also use the surface tension of water to cling to the surface and breathe air, through siphons. Surface tension decreases the ease with which gases dissolve in water.

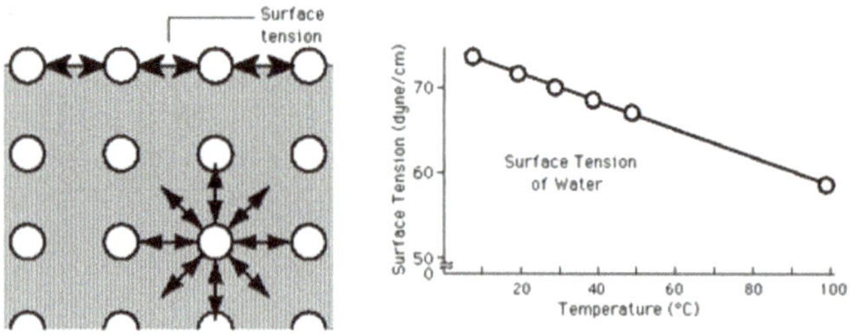

Figure 1.9: Hydrogen bonding results in surface tension

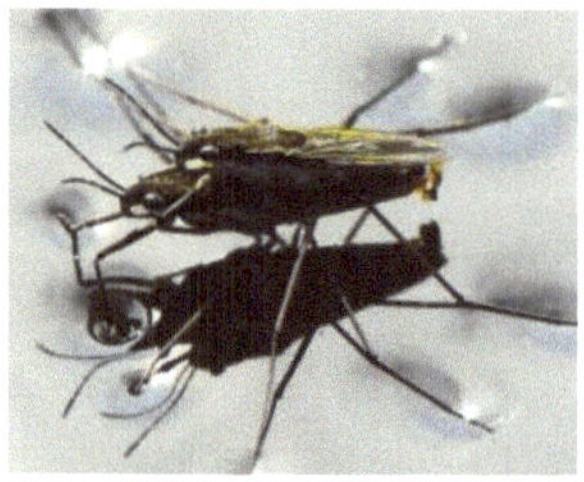

Pond Skater

Mosquito larva and pupa

8. *Water has maximum density at 4^0 C.* This means that the densest water at 4^0 C will remain at the bottom of an aquatic habitat. This prevents aquatic habitats from freezing completely, so that aquatic organisms can survive at the bottom, unfrozen at 4^0 C. Ice floats because maximum Hydrogen bonding occurs at 0^0 C. The ice on the surface of a frozen lake will serve as **an insulating layer** and reduce heat loss from the water body, so that the rest of the water at the bottom does not freeze.

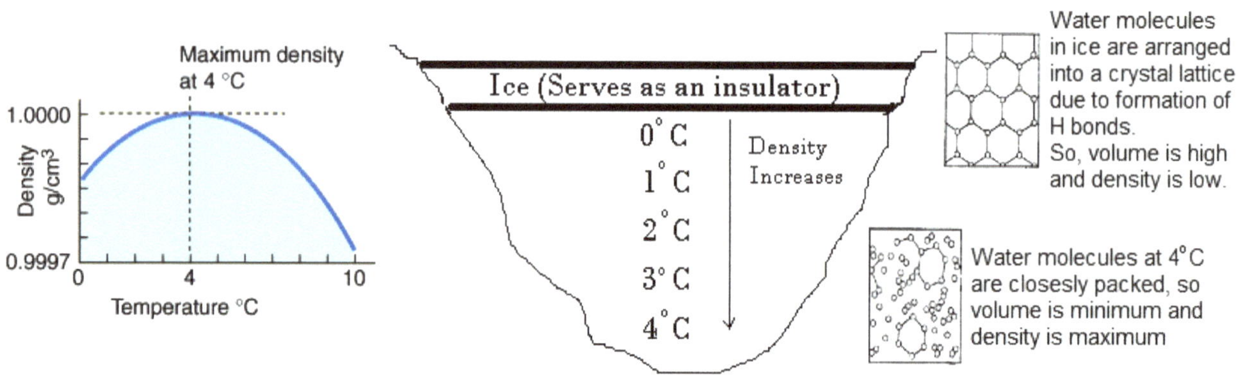

Figure 1.10: Temperature gradient in a frozen pond

Self-assessment (Chapter 1)

Step 1: Visit www.teachingtube.com

Step 2: Sign up for the free course "Edexcel AS and AL Bio assessments"

Step 3: Attempt the chapter Quiz and get instant results.

Step 4: Download the assessment worksheet and marking scheme too.

CHAPTER TWO
CARBOHYDRATES

Edexcel Specification 1.2: (i) know the difference between monosaccharides, disaccharides and polysaccharides, including glycogen and starch (amylose and amylopectin)
(ii) be able to relate the structures of monosaccharides, disaccharides and polysaccharides to their roles in providing and storing energy.
β-glucose and cellulose are not required in this topic.

Carbohydrates are polyhydroxy aldehydes or ketones, containing Carbon, Hydrogen and Oxygen only. Figure 2.1 shows the general classification of carbohydrates.

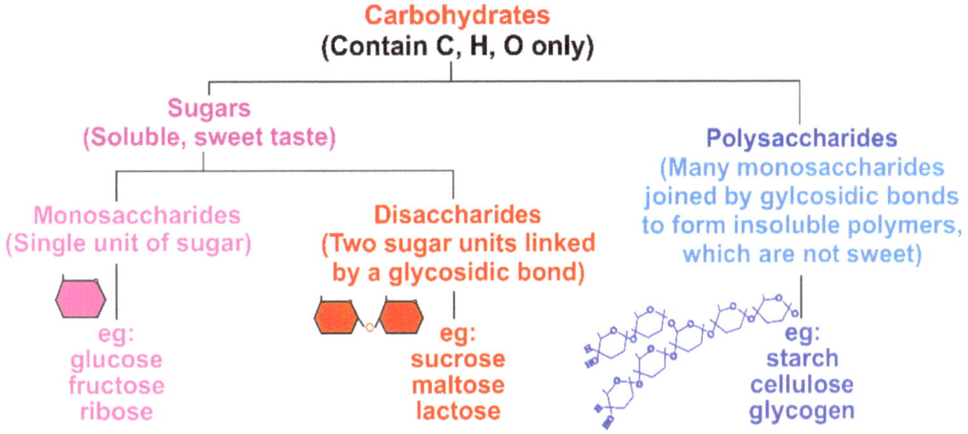

Figure 2.1: Types of carbohydrates

Differences between monosaccharides, disaccharides and polysaccharides

Monosaccharides	*Disaccharides*	*Polysaccharides*
Made up of single units of sugar with no glycosidic bond.	Made up of two monosaccharides linked by a single glycosidic bond.	Made up of many monosaccharides linked by glycosidic bonds.
Soluble in water.	Soluble in water.	Usually insoluble in water.
Sweet taste.	Sweet taste.	Not sweet.
Genera formula: $(CH_2O)_n$, where n is the number of Carbon atoms.	Genera formula: $2[(CH_2O)_n] – H_2O$, where n is the number of Carbon atoms.	Genera formula: $X[((CH_2O)_n) – ((X\text{-}1)\ H_2O)]$, where n is the number of Carbon atoms and X is the number of monosaccharide residues.

Monosaccharides: Energy transporters and providers

The table below refers to the features and functional utility of monosaccharides.

Feature	*Functional utility*
Polar and soluble	Glucose is transported in the blood of mammals as it is polar and soluble. It can easily enter the cells and get metabolized or stored as starch or glycogen.
Smaller and more reactive when compared to disaccharides and polysaccharides	Monosaccharides are the smallest and most reactive when compared to disaccharides and polysaccharides. Hence, they can easily enter the cells and be broken down by respiration to **provide energy for metabolic processes**. **Glucose + oxygen → Carbon dioxide + Water + Energy (ATP)**
Ability to undergo condensation reactions and form glycosidic bonds	Monosaccharides are used as building blocks for larger and more complex carbohydrates, like starch, glycogen and cellulose. Two monosaccharides may also be joined together by a glycosidic bond to form a disaccharide.

Disaccharides: Energy transporters and storage compounds

The table below refers to the features and functional utility of disaccharides.

Feature	*Functional utility*
Polar and soluble	Since disaccharides are polar they can dissolve in water. This is particularly useful for the transport of carbohydrates, as sucrose, in the phloem sap of plants.
Ability to be hydrolysed	Sucrose is hydrolysed in the gut of mammals to form sucrose and fructose. The glucose is then absorbed into the cells and used as a respiratory substrate. **Sucrose → glucose + fructose** **Glucose + oxygen → Carbon dioxide + Water + Energy (ATP)**
Less permeable to cell membranes	Disaccharides are larger than monosaccharides and cannot easily pass across the cell membranes. Hence, some plants like sugarcane store carbohydrates as sucrose.

Polysaccharides: Energy storage compounds

The polysaccharides that store energy are starch (in plants) and glycogen (in animals and bacteria). Cellulose is a component of the cell wall and plays no role in energy storage. The table below refers to the features and functional utility of starch and glycogen.

Feature	**Functional utility**
Insoluble	Polysaccharides are insoluble and cannot leave the cell.
Insoluble	Polysaccharides are insoluble and have no osmotic effect. So, cells can store lots of polysaccharides without the need for osmoregulation.
Hydrolysed by enzymes	Polysaccharides can easily be **hydrolyzed by enzymes** into glucose and used for respiration in cells.
Relatively unreactive	Polysaccharides are relatively unreactive when compared to monosaccharides and disaccharides. They do not get involved in other metabolic reactions in the cell.
Compact	This means that a lot of glucose is stored in a small space within the cell.

Edexcel Specification 1.3: CORE PRACTICAL 1
Use a semi-quantitative method with Benedict's reagent to estimate the concentrations of reducing sugars and with iodine solution to estimate the concentrations of starch, using colour standards.

Semi-quantitative test for estimation of reducing sugars

- *Prepare a range of reducing sugar solutions with concentrations of 0%, 0.1%, 0.5%, 1.0 %, and 5.0%.*
- *Equal volumes of the 0.1% sugar solution and Benedict's solution is pipetted into a test tube and boiled in a water bath, as shown in figure 2.2. The procedure is repeated with all the other sugar solutions.*
- *The colour of the solutions will change, as shown in figure 2.2. These are known as colour standards.*

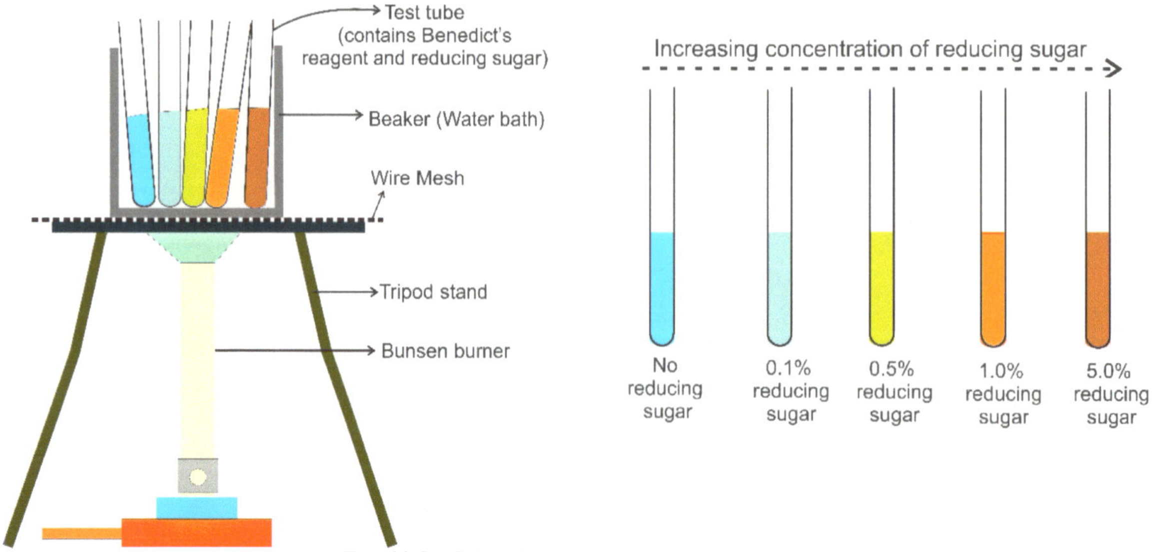

Figure 2.2: Benedict's test for reducing sugars (Semi-quantitative)

- ***The colour of the solutions change because reducing sugars*** *will react with Cu^{2+} ions in Benedict's solution and reduce the blue coloured Cu^{2+} ions into* ***brick-red colour Cu^{+} ions****.*
- *The colour changes from blue to green to yellow and finally to red, as the concentration of the red Cu^{+} ions increase.*
- *Solutions with low concentrations of reducing sugars will turn green or yellow and solutions with high concentrations of reducing sugar will turn orange or red, as shown in figure 2.2.*
- *Repeat the procedure with a reducing sugar solution of unknown concentration, or, with a food extract. Compare the colour of this solution with the colour standards to estimate the reducing sugar concentration in the unknown solution.*

- **Non-reducing sugars** will not change the colour of Benedict's solution. However, after acid or enzyme hydrolysis, it will change the colour of Benedict's solutions.
- **Acid hydrolysis:** Add hydrochloric acid to the sugar solution and heat it gently. Neutralize the mixture by adding Sodium Hydrogen carbonate solution or sodium hydroxide solution. Carry out the Benedict's test as explained above.
- **Enzyme hydrolysis:** The non-reducing sugar solution must be treated with hydrolytic enzymes and incubated. Then Benedict's test can be carried out. **Sucrose** is an example of a nonreducing sugar. It can be treated with sucrase or invertase enzyme to form glucose and fructose, which are reducing sugars.

Semi-quantitative test for estimation of starch

- *Prepare a range of starch solutions with concentrations of 0%, 0.01%, 0.1 %, and 1.0%.*
- *Equal volumes of the 0.01% starch solution and a solution of iodine in potassium iodide is pipetted into a test tube.*
- *The procedure is repeated with all the other solutions.*
- *The colour of the solutions will change, as shown in figure 2.3.*
- *These are known as colour standards.*

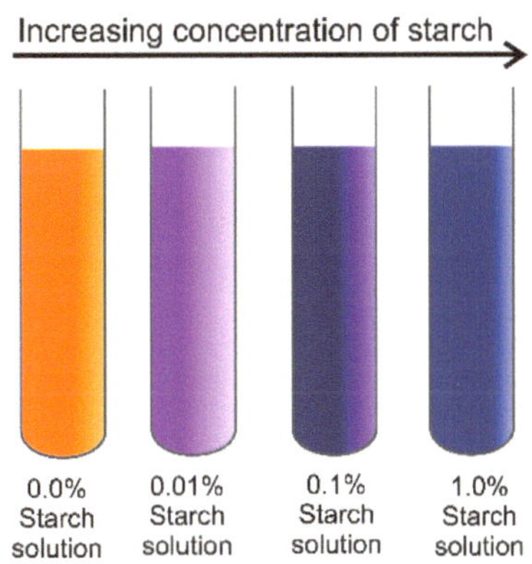

Figure 2.3: Iodine test for starch (Semi-quantitative)

- *The colour of the solution changes because iodine gets trapped in the core of the amylose helix of starch to form a blue-black complex.*

- *Repeat the procedure with a starch solution, or, food extract of unknown concentration.*
- *Compare the colour of this solution with the colour standards to estimate the starch concentration in the unknown solution.*

Edexcel Specification 1.4: know how monosaccharides (glucose, fructose and galactose) join together to form disaccharides (maltose, sucrose and lactose) and polysaccharides (glycogen, amylose and amylopectin) through condensation reactions forming glycosidic bonds, and how these can be split through hydrolysis reactions.

Condensation reaction

During a condensation reaction, two monosaccharides are joined by the **removal of a water molecule** to form a **glycosidic bond**. Condensation is brought about by specific enzymes.

Maltose formation by a condensation reaction

Maltose is commonly called malt sugar. It is an essential ingredient in beer production.

It consists of two α glucose residues linked by a 1, 4 – glycosidic bond, formed by a condensation reaction as shown in figure 2.4. Maltose is an intermediate product of starch hydrolysis in plants and germinating seeds.

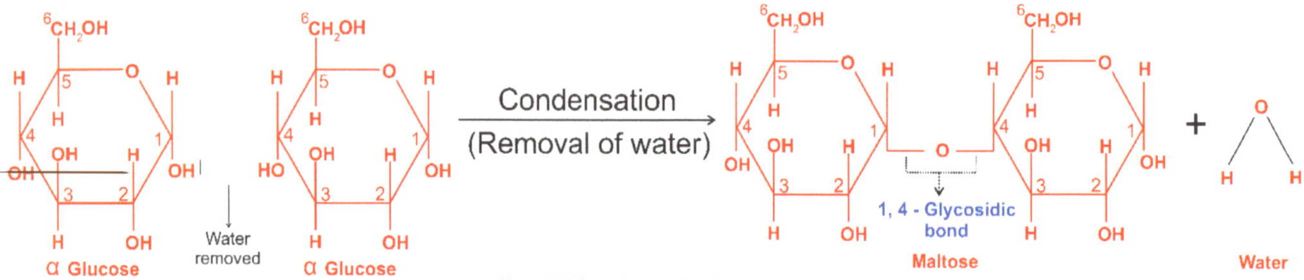

Figure 2.4: Formation of maltose by condensation

Lactose formation by a condensation reaction

Lactose is commonly called milk sugar. It consists of a β galactose residue linked to an α glucose residue by a 1, 4 – glycosidic bond formed by a condensation reaction as shown in figure 2.5. Lactose is the sugar found in mammalian milk. It serves as a carbohydrate source, which provides energy to young mammals.

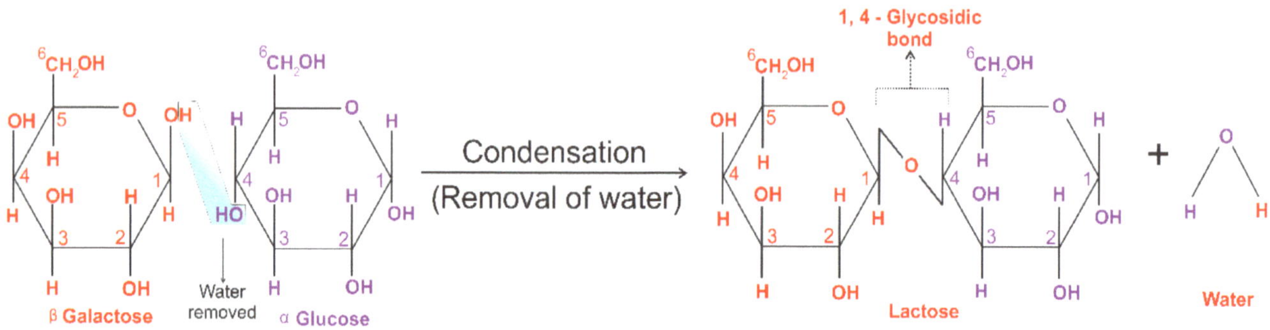

Figure 2.5: Formation of lactose by condensation

Sucrose formation by a condensation reaction

Sucrose is the common sugar that we use in puddings, tea and coffee. It consists of an α glucose residue linked to fructose by a 1, 2 – glycosidic bond formed by a condensation reaction as shown in figure 2.6. Sucrose is the form in which plants transport carbohydrates, through the phloem. Sucrose is also stored in the stems and roots of sugarcane and sugar beet plants.

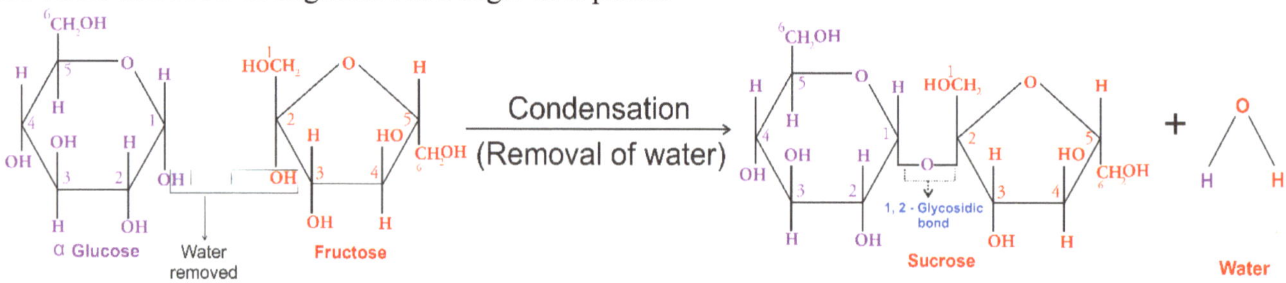

Figure 2.6: Formation of sucrose by condensation

Hydrolysis

Hydrolysis is the splitting of a disaccharide into its monosaccharides, by the **addition of a water molecule**.

Hydrolysis can be brought about by

- hydrolase enzymes or
- by treating with hydrochloric acid, followed by heating. The excess acid is neutralized by sodium hydrogen carbonate solution till the solution stops fizzing.

Hydrolysis of Maltose into two glucose molecules.

Maltose is hydrolysed into 2 alpha glucose molecules by the addition of water. The enzyme maltase hydrolyses the 1,4 – glycosidic bond. The reaction is shown in figure 2.7.

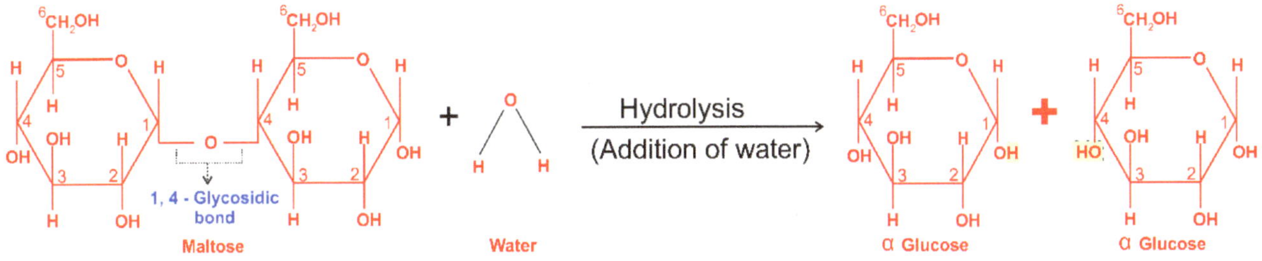

Figure 2.7: Hydrolysis of maltose

Hydrolysis of Lactose into glucose and galactose.

Lactose is hydrolysed into one alpha glucose molecule and one beta galactose molecule by the addition of water. The enzyme lactase hydrolyses the 1,4 – glycosidic bond. The reaction is shown in figure 2.8.

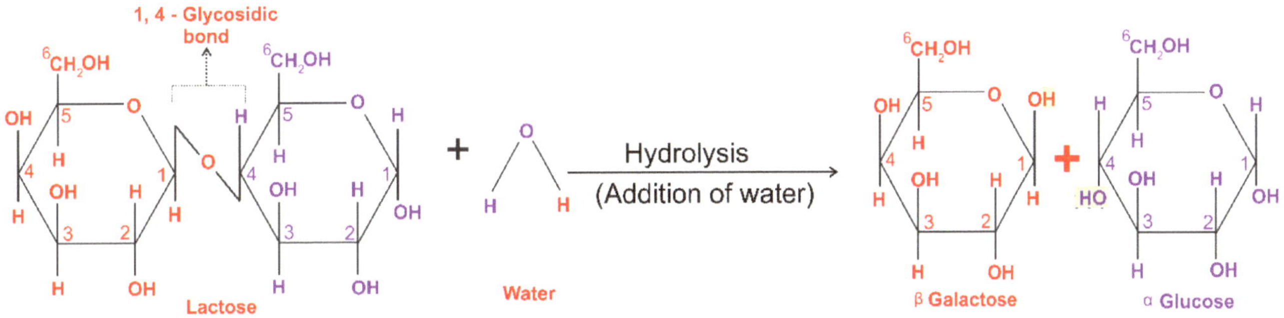

Figure 2.8: Hydrolysis of Lactose

Hydrolysis of Sucrose into glucose and fructose.

Sucrose is hydrolysed into one alpha glucose molecule and one fructose molecule by the addition of water. The enzyme sucrase or invertase hydrolyses the 1,2 – glycosidic bond. The reaction is shown in figure 2.9.

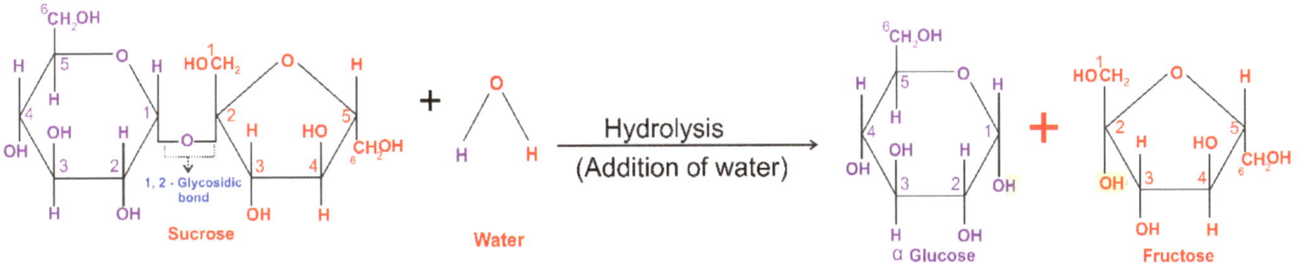

Figure 2.9: Hydrolysis of sucrose

Polysaccharides - Polymers

Many monosaccharides can be linked by glycosidic bonds to form polysaccharides.

Starch

Starch is a polysaccharide made up of many α glucose residues linked by glycosidic bonds. Starch is a **mixture** of **amylose** and **amylopectin**. The main function of starch is that it is an **energy storage molecule** in plant cells.

Amylose structure: Starch is a mixture of **amylose** and amylopectin. Amylose is composed of α glucose residues linked by 1, 4 – glycosidic bonds, formed by condensation reactions. The unbranched chain is then coiled into an amylose helix due to formation of hydrogen bonds between the glucose residues, as shown in figure 2.10. There are six glucose residues per turn of the helix. The amylose reacts with iodine to form a blue-black complex, which is used as a common test for starch. Amylose can be hydrolysed by the enzyme amylase, to yield maltose.

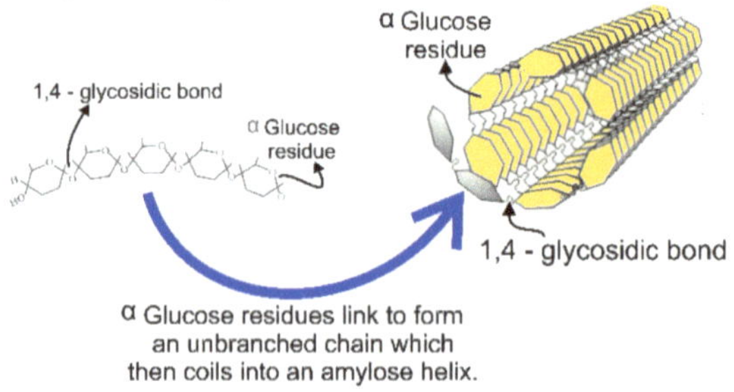

Figure 2.10: Formation of the amylose helix

Amylopectin structure: Amylopectin is also made up of α glucose residues. However, the glucose residues form a branched chain. The branches are formed due to the presence of 1, 6 – glycosidic bonds. The 1, 4 – glycosidic bonds are found in the unbranched part of the chain, as shown in figure 2.11. Branches usually form after every 20 to 30 glucose residues. Hydrolysis of starch is brought about by amylase. Slow release starch is found in potatoes and seeds. Amylopectin is broken down more rapidly than amylose as amylopectin is branched and many amylase molecules can act at the same time to remove maltose from the tip of each branch.

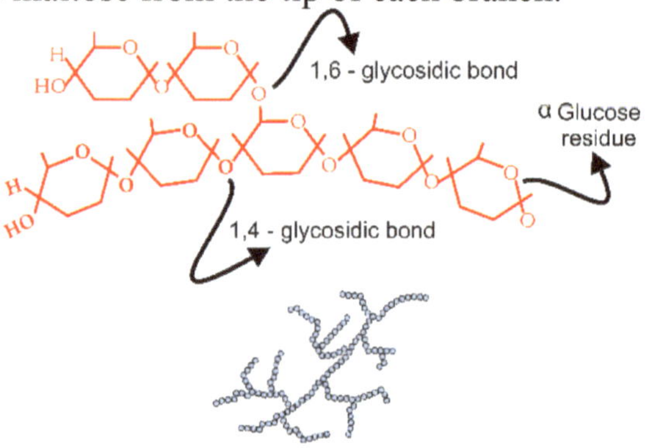

Figure 2.11: Formation of amylopectin

Glycogen

Glycogen is also made up of α glucose residues. The glucose residues form a branched chain. The branches are formed due to the presence of 1, 6 – glycosidic bonds. The 1, 4 – glycosidic bonds are found in the unbranched part of the chain. Branches usually form after every 8 to 10 glucose residues. Glycogen synthase is an enzyme involved in the formation of glycogen, by condensation reactions. Glycogen phosphorylase is the primary enzyme involved in the hydrolysis of glycogen. The glycogen is **stored** in the liver and muscle tissues of animals and can readily be hydrolysed to provide glucose for cellular respiration. The structure of glycogen and amylopectin are similar, but glycogen is more branched, as shown in figure 2.12.

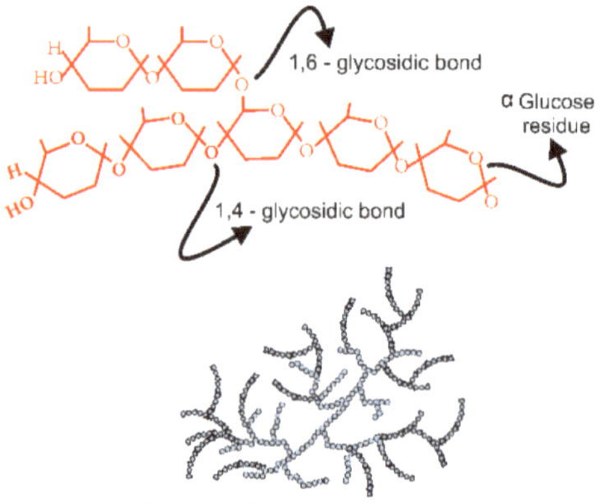

Figure 2.12: Formation of amylopectin

CHAPTER THREE

LIPIDS

Edexcel IAL from 2018 Specification 1.5: (i) know how a triglyceride is synthesised by the formation of ester bonds during condensation reactions between glycerol and three fatty acids
(ii) know the differences between saturated and unsaturated lipids

Lipids are a large group of organic compounds made up of Carbon, Hydrogen and Oxygen. They are **polyesters** formed by linking of **glycerol** with **fatty acid chains** by **condensation reactions**. Fats and oils are chemically similar, but fats are solids and oils are liquid at room temperature. Waxes have long chained hydrocarbons linked to their fatty acids. All lipids are **nonpolar**, hence **insoluble in water.** The figure 3.1 shows the different types of lipids.

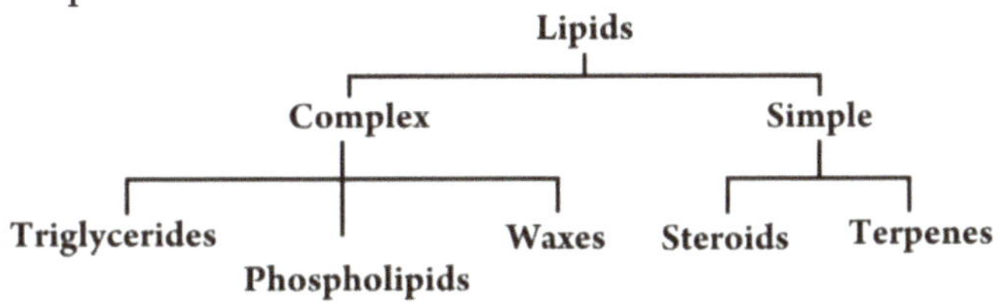

Figure 3.1: Types of lipids

Lipids have various roles in living things.

- The cell membranes are made of phospholipids.
- Adipose tissue below the skin stores energy as fats and provides thermal insulation and buoyancy in aquatic mammals.
- Sebum is an oily secretion which waterproofs the skin.
- Waxes are used to make beehives.
- Oil is used to preen the feathers of aquatic birds. Steroids are used to make hormones.

Triglycerides

Triglycerides are the most common form of lipids. It consists of one glycerol molecule and three fatty acid chains. The hydroxyl (-OH) groups of glycerol and the carboxylic acid (-COOH) group of each fatty acid are linked by an ester bond formed by condensation reactions, as shown in figure 3.2.

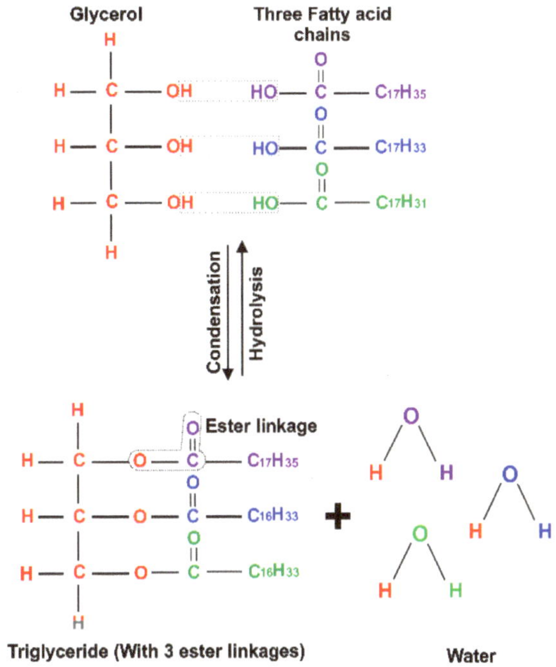

Figure 3.2: Formation of triglycerides

A triglyceride comprises of three fatty acid chains linked to a glycerol. Each fatty acid comprises of a carboxylic acid group (-COOH) linked to a hydrocarbon tail, as shown in figure 3.3. The triglyceride can have all three saturated fatty acid chains or a combination of saturated and unsaturated fatty acid chains attached to the glycerol residue. The length of the hydrocarbon tail also determines the properties of the triglyceride.

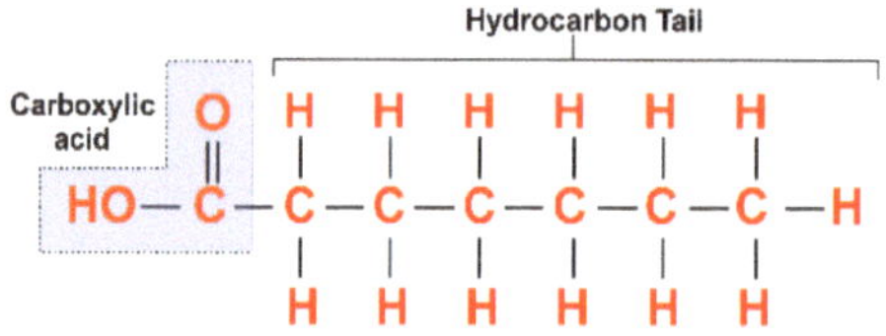

Figure 3.3: Structure of a fatty acid

Fatty acids may be saturated fatty acids or unsaturated fatty acids. The difference between saturated and unsaturated fatty acids is discussed below.

Saturated fatty acids	Unsaturated fatty acids
Have only single bonds between carbon atoms (C-C) in the hydrocarbon chain, as shown in figure 3.4. 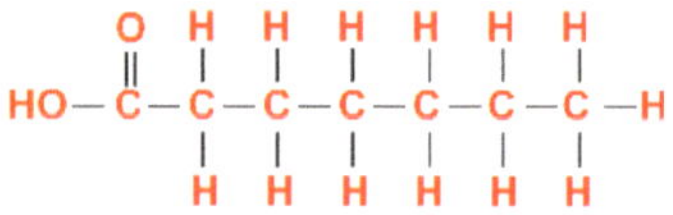 *Figure 3.4: Structure of a saturated fatty acid*	Have at least one double bond (- C = C -) **between carbon atoms** in the hydrocarbon chain, as shown in figure 3.5. 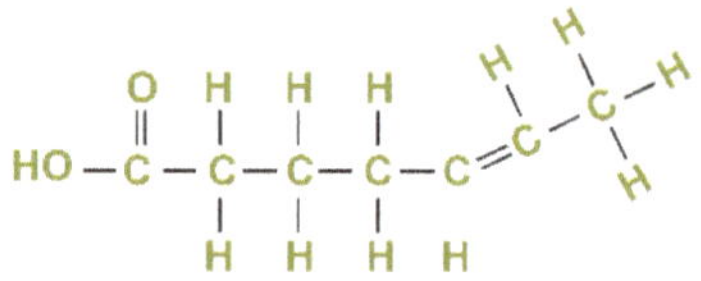 *Figure 3.5: Structure of an unsaturated fatty acid*
These fatty acids form straight chains, as shown in figure 3.4.	These fatty acids form **bent** chains. The chains have kinks where double bonds are present, as shown in figure 3.5.
Have a high melting point. Hence, they are usually solids at room temperature. Example, Fats such as butter and Stearic acid ($CH_3(CH_2)_{16}COOH$).	Have a low melting point. Hence, they are usually liquids at room temperature. **Examples:** oleic acid – MUFA, linoleic acid - PUFA.
Have the general formula $\mathbf{C_nH_{2n+1}COOH}$. Saturated fatty acids have more number of Hydrogen atoms than an unsaturated hydrocarbon with the same number of Carbon atoms. For example, Stearic acid has the formula $\mathbf{C_{16}H_{35}COOH}$. It has more hydrogen atoms than oleic and linoleic acid which are unsaturated fatty acids with the same number of carbon atoms.	Fatty acids with one double bond are called Monounsaturated fatty acids (MUFAs) and have the general formula $\mathbf{C_nH_{2n-1}COOH}$. Fatty acids with more than one double bond are called polyunsaturated fatty acids (PUFAs) and have the general formula $\mathbf{C_nH_{2n-3}COOH}$. So, Unsaturated fatty acids have lesser number of Hydrogen atoms than a saturated hydrocarbon with the same number of Carbon atoms. For example, oleic acid has the formula $\mathbf{C_{16}H_{33}COOH}$ and linoleic acid has the formula $\mathbf{C_{16}H_{31}COOH}$. Both are unsaturated fatty acids with 16 Carbons each and have fewer hydrogen atoms than stearic acid.

Self-assessment (Chapter 3)

Step 1: Visit www.teachingtube.com

Step 2: Sign up for the free course "Edexcel AS and AL Bio assessments"

Step 3: Attempt the chapter Quiz and get instant results.

Step 4: Download the assessment worksheet and marking scheme too.

Chapter Four

The circulatory system

Edexcel IAL from 2018 Specification 1.6: understand why many animals have a heart and circulation (mass transport to overcome the limitations of diffusion in meeting the requirements of organisms)

The need for a circulatory system

Single celled organisms, such as amoeba, exchange food, respiratory gases and excretory products from the environment, directly from the cell surface membrane. They have a large surface area to volume ratio and a low diffusion distance. So, diffusion from the body surface alone is enough to sustain metabolism, as shown in figure 4.1.

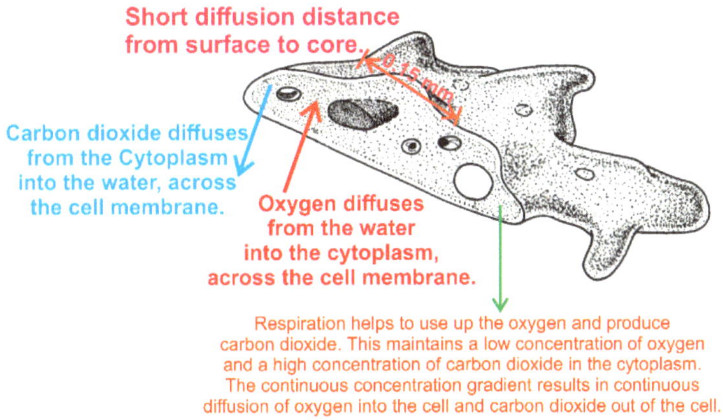

Figure 4.1: Transport of gases in amoeba

However, animals are multicellular and have a very **low surface area to volume ratio** and **high diffusion distance** from the surface to the body core. **Diffusion from the body surface would be too slow to supply or remove materials from the cells at a suitable rate to sustain metabolism in these cells**. So, multicellular organisms need a circulatory system to supply or remove materials from the cells at a rapid rate.

One of the main functions of the circulatory system in mammals is the transport of respiratory gases, metabolites, metabolic wastes, and hormones. As organisms get bigger their surface area to volume ratio gets smaller, as shown in figure 4.2. As organisms become bigger it is more difficult for them to rapidly transport materials from their surface to the core. Therefore, exchange of materials from the surfaces alone is not efficient enough to sustain life.

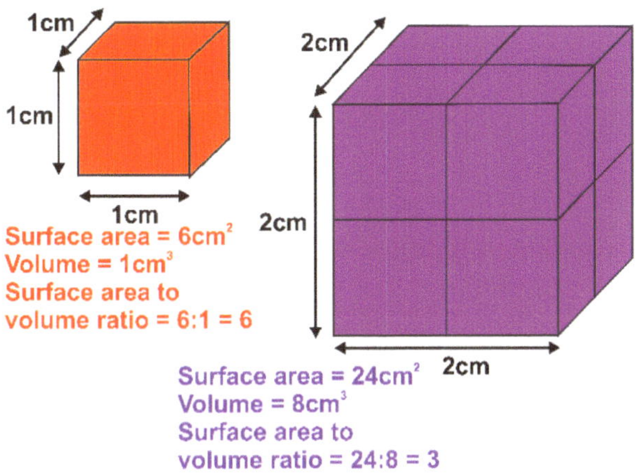

Figure 4.2: Relationship between size and surface area to volume ratio

Blood vessels

The human circulatory system is a closed circulatory system. The blood does not leave the blood vessels. The blood is pumped by the heart into arteries and brought back to the heart by veins. The capillaries connect arteries to veins.

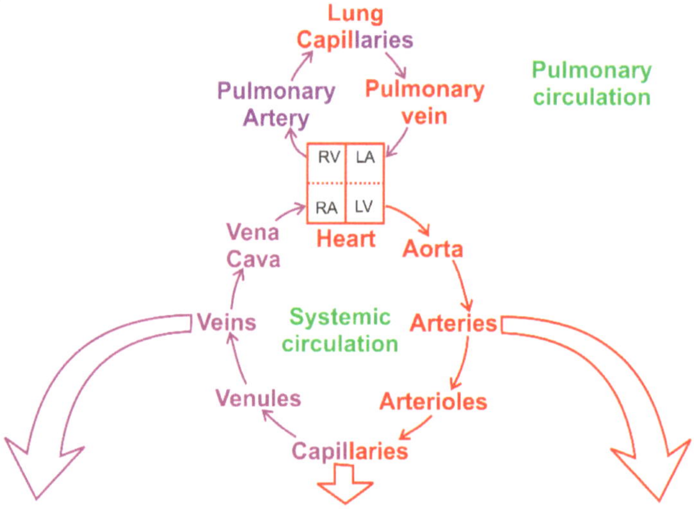

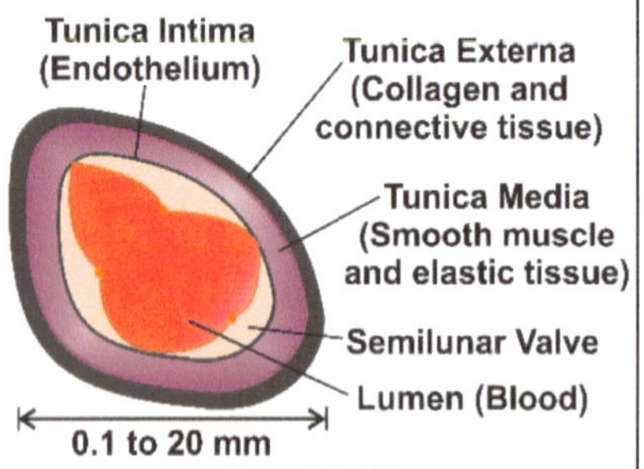

Figure 4.3: Vein

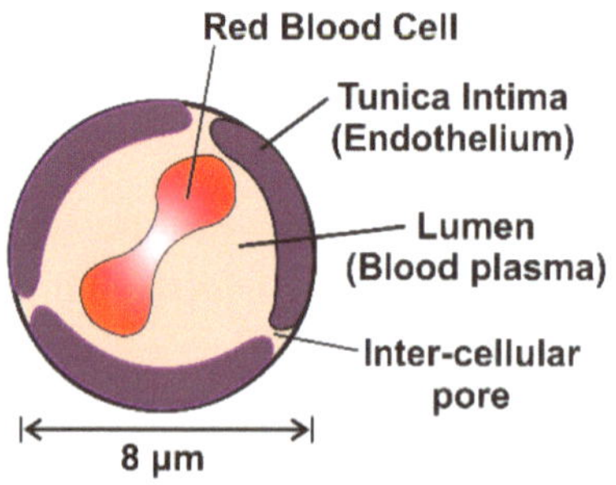

Figure 4.4: Capillary

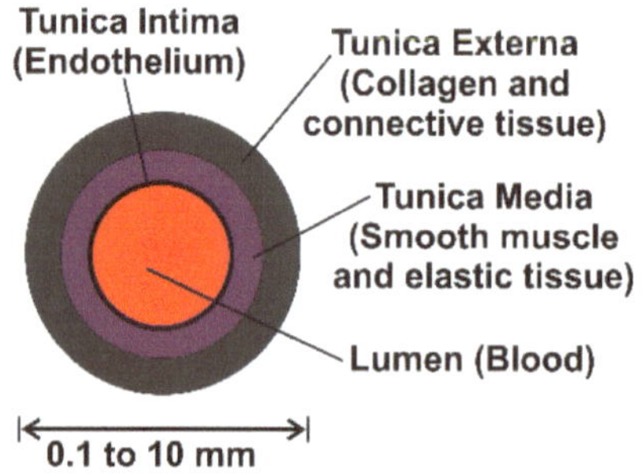

Figure 4.5: Artery

Veins and Venules	Capillaries	Arteries and Arterioles
Function is to carry blood from tissues towards the heart.	Function is to allow exchange of materials between the blood and the tissues.	Function is to carry blood from the heart towards the tissues.
Thin walls, mainly collagen, since blood at low pressure.	Very thin, permeable walls, only one cell thick to allow easy diffusion of materials.	Thick walls with smooth elastic layers to resist high pressure and muscle layer to aid pumping (recoil).
Large lumen to reduce resistance to blood flow.	Very small lumen. Blood cells must distort to pass through.	Small lumen contributes to the high pressure.
Many valves to prevent backflow of blood.	No valves.	No valves (except in the heart at the origin of the pulmonary artery and the aorta).
Blood at low pressure due to large lumen.	Blood pressure falls in capillaries as volume of blood decreases due to formation of tissue fluid and the large surface area of the capillary network.	Blood at high pressure due to narrow lumen and elastic recoil of the artery walls.
Blood usually deoxygenated (except in pulmonary vein and umbilical vein).	Blood changes from oxygenated to deoxygenated (except in lungs).	Blood usually oxygenated (except in pulmonary artery and umbilical artery).

- **Collagen** makes vessels strong and durable.
- **Elastic fibres** allow vessels to stretch and recoil.
- **Smooth muscle cells** in the walls allow vessels to constrict and dilate.

Blood flow in veins.

The **skeletal muscles contract to propel the blood** in the veins towards the heart. The **semi-lunar valves prevent the backflow of blood** in the veins, as in figure 4.6. **Breathing movements reduce the pressure in the thoracic cavity** and cause **a suction force** to assist the flow of blood towards the heart.

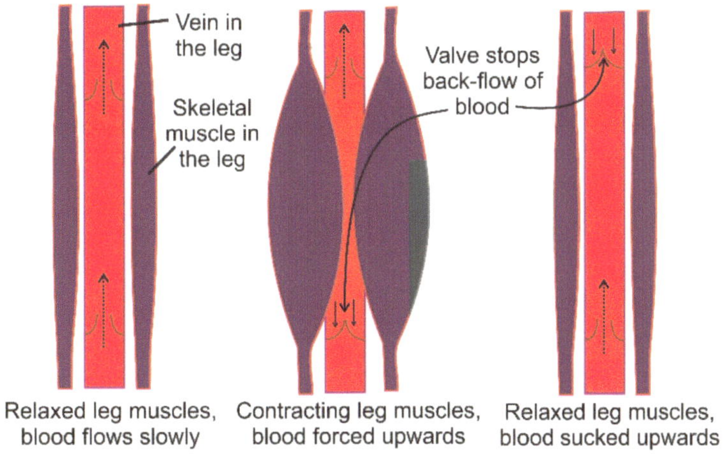

Figure 4.6: Transport of blood in the veins

The body relies on constant contraction of skeletal muscles to get the blood back to the heart. This explains why soldiers standing still on parade for long periods can faint, and why sitting still on a long flight can cause swelling of the ankles and Deep Vein Thrombosis (DVT or "economy class syndrome"), where small blood clots collect in the legs.

> Edexcel IAL from 2018 Specification 1.8: know the cardiac cycle (atrial systole, ventricular systole and cardiac diastole) and relate the structure and operation of the mammalian heart, including the major blood vessels, to its function
> *Details of myogenic stimulation not needed at IAS.*
>
> RECOMMENDED ADDITIONAL PRACTICAL
> Investigate the structure of a mammalian heart by dissection.

The Cardiac Cycle

The cardiac cycle is the rhythmic contraction and relaxation of the atria and ventricles during one complete heartbeat. The cardiac cycle can be divided in to three main stages, as shown in figure 4.7.

1) **Atrial systole, ventricular diastole (0.13 sec)** – The atria contract and push blood into the ventricles through atrio-ventricular valves.
2) **Ventricular systole, atrial diastole (0.3 sec)** – The ventricles contract and pumps blood from ventricles into arteries through semi-lunar valves. Atria are being refilled during this time.
3) **Complete diastole (0.4 sec)** – atria are being filled with blood from veins. Blood slowly oozes into ventricles from atria.

Systole means contraction;
Diastole means relaxation.

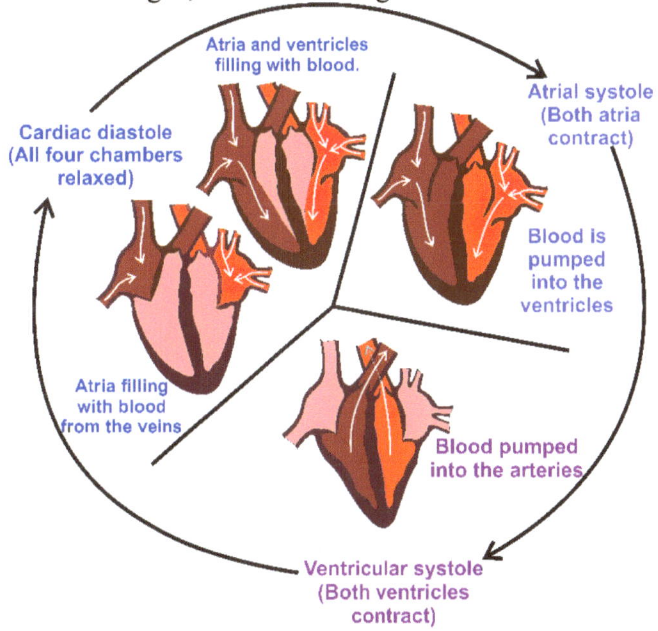

Figure 4.7: The cardiac cycle

Structure of mammalian heart

The mammalian heart is in the thoracic cavity. It is embraced by the lungs and slightly tilted to the left. That is why the powerful contraction of the left atrium can be felt on the left of the chest. The heart is enclosed in membranes called the pericardial membranes, which contain the pericardial fluid between them. The fluid absorbs mechanical shock.

The wall of the heart is made up three distinct layers:

- The outer epicardium – consisting of flattened epithelial cells and supporting connective tissue;
- a thick layer of myocardium – consisting of cardiac muscles cells; and
- an inner endocardium – made up of flattened epithelial cells and connective tissue.

Figure 4.8 shows the major structures in the heart.

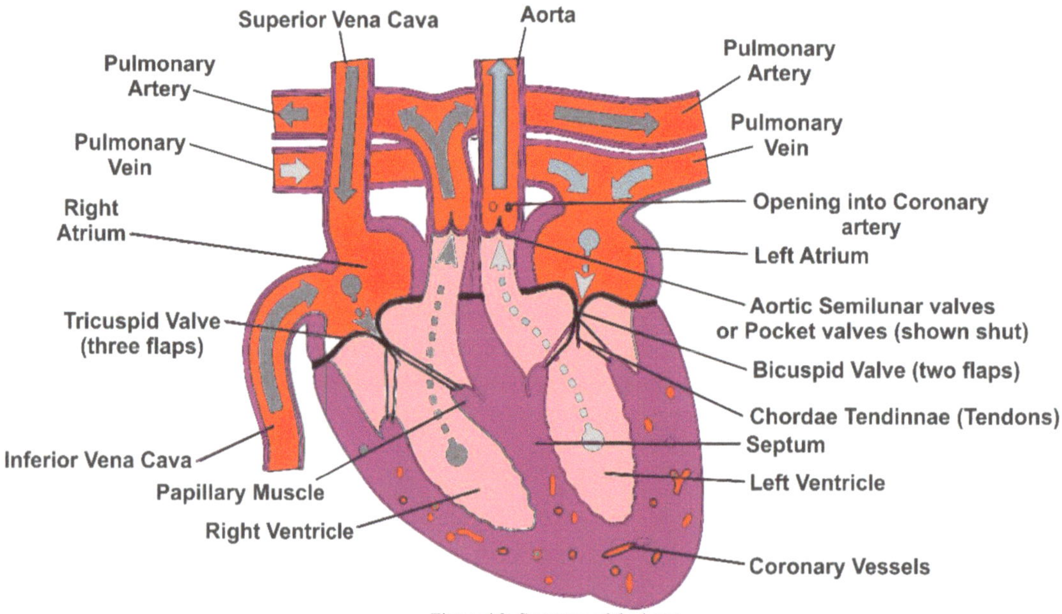

Figure 4.8: Structure of the heart

Structure	Function
Superior Vena Cava	Carries deoxygenated blood from the **head** to the **right atrium**.
Inferior Vena Cava	Carries deoxygenated blood from the **body** to the **right atrium**.
Aorta	Carries oxygenated blood from the **left ventricle** to the **body**.
Pulmonary artery	Carries deoxygenated blood from the **right ventricle** to the **lungs**.
Pulmonary vein	Carries oxygenated blood from the **lungs** to the **left atrium**.
Left Atrium	Receives oxygenated blood from pulmonary vein and pumps it into the left ventricle
Right Atrium	Receives deoxygenated blood from vena cava and pumps it into the right ventricle
Left Ventricle	Receives oxygenated blood from the left atrium and pumps it to all the organ systems at high pressure (systemic circulation)
Right Ventricle	Receives deoxygenated blood from the right atrium and pumps it to the lungs at lower pressure (pulmonary circulation)
Pocket valves or semilunar valves	Prevents the backflow of blood from the arteries into the ventricles.
Bicuspid and tricuspid valve (atrio-ventricular valves)	Prevents backflow of blood into the atria during ventricular systole. The closing of these valves also helps to build up a high pressure in the ventricles during ventricular systole, which helps to push blood into the arteries.
Coronary artery	Supplies oxygenated blood to the cardiac muscle
Papillary muscles	Maintains the tension on the *chordae tendinnae*
Chordae tendinnae (tendons)	Prevents the bicuspid and tricuspid valves from flipping over into the atria during ventricular systole.
Septum	Separates the right and left sides of the heart, so that oxygenated blood in the left cannot mix with deoxygenated blood in the right side of the heart.

The walls of the left ventricles are much thicker as they must exert a greater pressure to pump blood to all parts of the body, even to the head against the force of gravity. The right ventricles exert a much lower pressure as the lungs are close to the heart and pulmonary tissues are very delicate. High pressure in the pulmonary circulation will lead to pulmonary oedema, causing the lungs to fill with tissue fluid. This will cause the person to drown in their own tissue fluid. The left side of the heart contains only oxygenated blood and the right side contains only deoxygenated blood. The right and left sides are separated by a muscular wall called as the septum.

Double circulatory system

The mammalian circulatory system consists of two circulations, as shown in figure 4.9.
Systemic circulation is the pumping of oxygenated blood to all the organ system of the body. It is a *high-pressure circulation* originating from the left ventricle and terminating at the right atrium, which receives deoxygenated blood from the organ systems or tissues.

Pulmonary circulation is the pumping of deoxygenated blood to the lungs. It is a *low-pressure circulation* originating from the right ventricle and terminating at the left atrium, which receives oxygenated blood from the lungs.

Double circulation ensures that oxygenated blood does not mix with deoxygenated blood. So, only oxygenated blood is circulated to all the organ systems of the body.

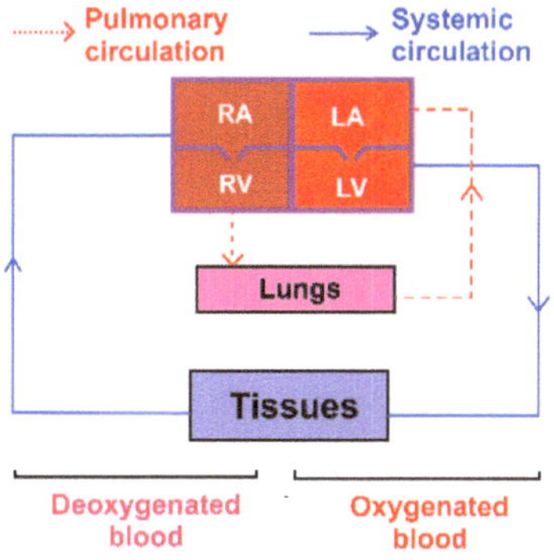

Figure 4.8: Double circulation

The Coronary circulation

The Coronary circulation supplies blood to the cardiac muscles. The coronary artery branches from the aorta and enters the heart muscles, as shown in figure 4.9. It splits into capillaries and exchange of materials takes place between the blood and tissues. The capillaries reunite to form cardiac veins, which pour blood back into the right atrium through the coronary sinus, as shown in figure 4.10. About 5% of the total cardiac output goes to coronary circulation. Blockage of the coronary vessels leads to heart attacks or Coronary heart diseases (CHD).

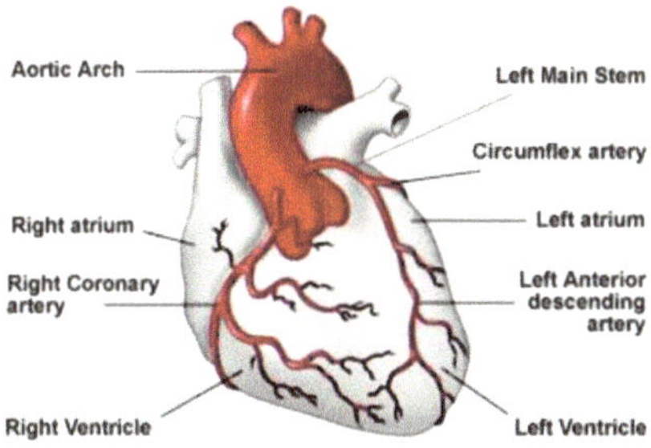

Figure 4.9: Coronary Arteries

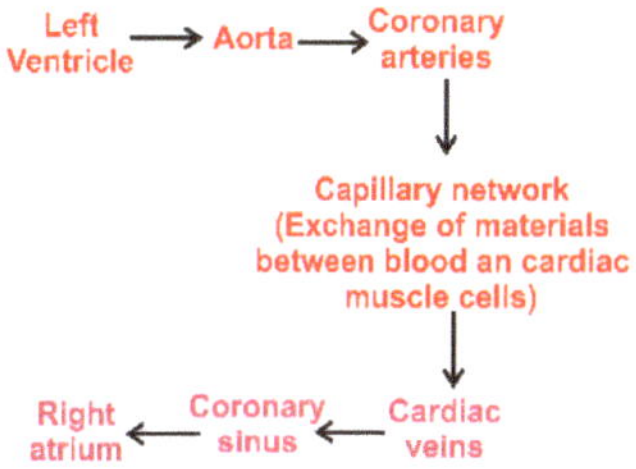

Figure 4.10: Pathway of blood in coronary circulation

The cardiac muscles contract and relax continuously throughout life. Muscle contraction requires large amounts of ATP, which is produced by respiration. So, the cardiac muscle needs a rapid and continuous supply of oxygenated blood to enable ATP production by aerobic respiration. This is made possible by the coronary circulation.

Lack of oxygen in cardiac muscle tissue, due to blocks in the coronary arteries, will result in anaerobic respiration, which produces lactic acid and very small quantities of ATP. The lactic acid will lower the pH inside the muscle cells and deactivates muscle proteins and enzymes, leading to muscle fatigue and cardiac arrest or heart attacks. The lack of ATP will also prevent the muscles from contracting.

The Figure 4.13 is a graph showing the pressure changes in the heart during a complete cardiac cycle. The letters on the graph show various events that occur during one complete cardiac cycle.

OPENING and CLOSING of semi-lunar valves

The semi-lunar valves are open whenever pressure in the ventricles is greater than pressure in the arteries, as shown in figure 4.11. At all other times semi-lunar valves are closed.

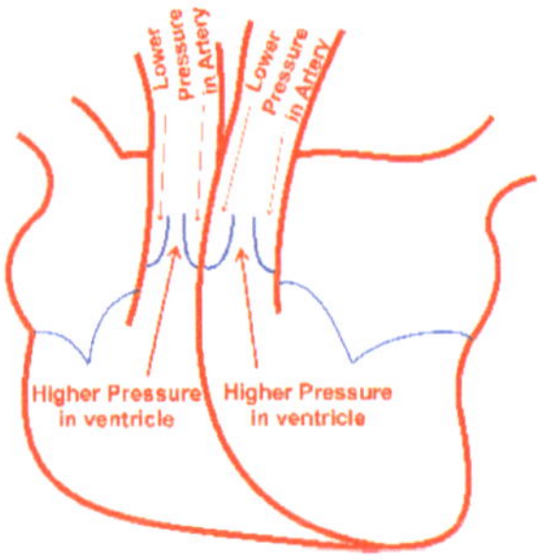

Figure 4.11: Opening of the semi-lunar valves

OPENING and CLOSING of atrio-ventricular valves

Atrio-ventricular valves are open whenever the pressure in atria is greater than pressure in ventricles as shown in figure 4.12. At all other times the atrio-ventricular valves are closed.

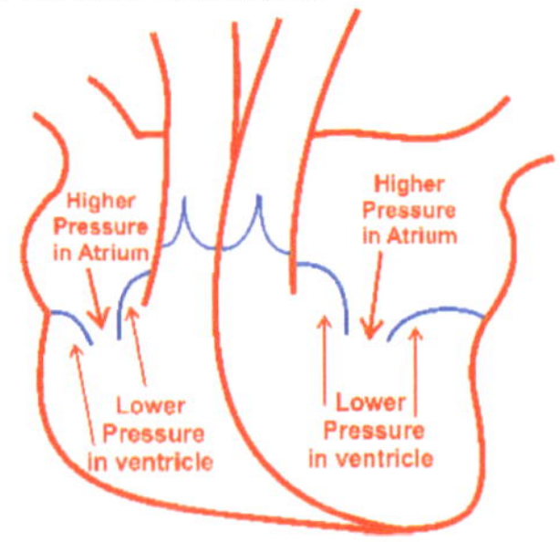

Figure 4.12: Opening of the Atrio-ventricular valves

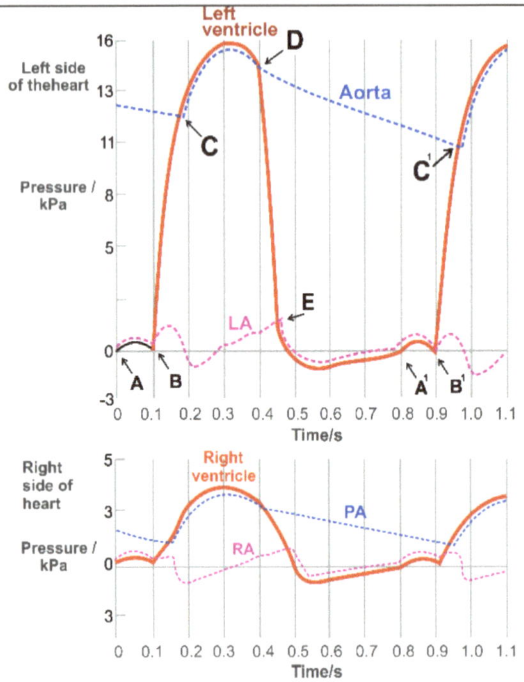

Figure 4.13: Pressure changes in the heart

The pressure changes in both sides of the heart occur simultaneously. Both ventricles contract together and both atria contract together. However, the pressure in the right side is lower due to the thinner walls of the right ventricle.

Elastic recoil of arteries: During ventricular systole, the aorta is stretched due to the high pressure. During complete diastole, the aortic semilunar valve closes, and the aorta walls recoil to maintain a high pressure. This pressure helps to propel the blood forward in the arteries. The elastic recoil of arteries can be felt as the pulse.

A – A^1: one complete heartbeat.

A – B: atrial systole (0 to 0.1s).

B – D: ventricular systole (0.1 to 0.4s).

D – A^1: complete diastole (0.4 to 0.8s).

at B: Ventricular systole begins. Atrio-ventricular valves close. **'Lubb'** sound.

B – C: Ventricular systole continues, but semi-lunar valves remain closed as pressure in aorta is greater than in ventricle.

at C: semi-lunar valves open as pressure in ventricle now exceeds pressure in aorta.

C – D: Ventricular systole continues, and blood is pumped into the aorta. Aorta walls are stretched due to the high pressure.

at D: Semi-lunar valves close with a **'Dubb'** sound. Pressure in ventricle is lower than in aorta as all the blood has been pumped out of ventricle and ventricles are relaxed.

D – C^1: Elastic recoil of aorta walls maintains a high pressure in the aorta and pushes the blood forward into the arteries.

at E: Atrio-ventricular valves open as blood in the atrium pushes the valve open and oozes or trickles into the ventricles.

E – A^1: blood keeps flowing into the ventricles from the atria.

Structure of mammalian heart related to its function

1. **The walls of the left ventricle are thicker than the right ventricle.** This ensures that there is a high pressure in the systemic circulation, as the left ventricle must pump blood to all parts of the body (even to the head, against the force of gravity). The right ventricle pumps blood to the lungs at a lower pressure, as the lungs are very close to the heart and pulmonary oedema would occur if the pressure is too high.

2. **The Sino Atrial Node (SAN) is located on the roof of the atrium**, so that the atria contract from the top of the heart and push blood downwards into the ventricles.

3. **The Purkinje fibres extend into the ventricle walls from the apex.** This ensures that ventricular systole begins from the apex and pushes blood upwards into the arteries. It also results in the closing of the atrio-ventricular valves.

4. **The heart muscle has its own blood supply.** Even though the heart is filled with blood, the walls of atria and ventricles have too small a surface area to allow diffusion of substances to or from all heart cells. The coronary circulation ensures that every heart muscle cell is close to a capillary for rapid diffusion of oxygen, glucose, nutrients, waste products and carbon dioxide. This helps the cardiac muscle to continue beating rapidly without getting exhausted.

5. **There is a non-conducting layer between the atria and ventricles.** The impulses can only pass to the ventricles via the AVN, thus causing a delay and ensuring coordinated contraction of the ventricles after the atria have completely contracted.

Coordination of heartbeat (Cardiac cycle)

Myogenic stimulation of the cardiac cycle means that the stimulus for the contraction of cardiac muscles is generated within the heart itself (not generated by brain or spinal cord). This is evident from the fact that isolated cardiac muscle cells will continue to contract and relax rhythmically when placed in an isotonic solution, without any electrical stimulation. This indicates that cardiac muscles are self-excitatory (myogenic). However, the rate of heartbeat can be altered by the nervous system or hormones (like adrenaline). Figure 4.14 shows the conduction system of the heart.

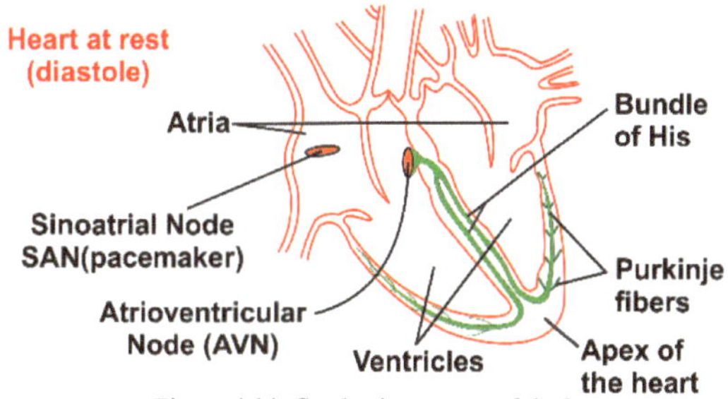

Figure 4.14: Conducting system of the heart

The impulse for heartbeat originates at the Sino atrial node (SAN), also known as the pacemaker. This impulse (wave of excitation) spreads through the atrial muscles causing atrial systole, as shown in figure 4.15. Intercalated discs of cardiac muscles enhance the transmission of excitation.

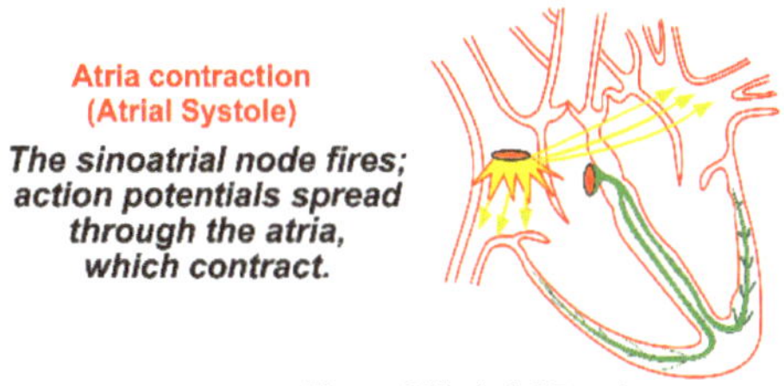

Figure 4.15: Atrial Systole

The impulse is then taken up by the Atrio-Ventricular Node (AVN) and passed down to the apex of the heart, by the bundle of His. The wave of excitation then spreads into the ventricular walls (through the Purkinje fibres) causing ventricular systole, as shown in figure 4.16.

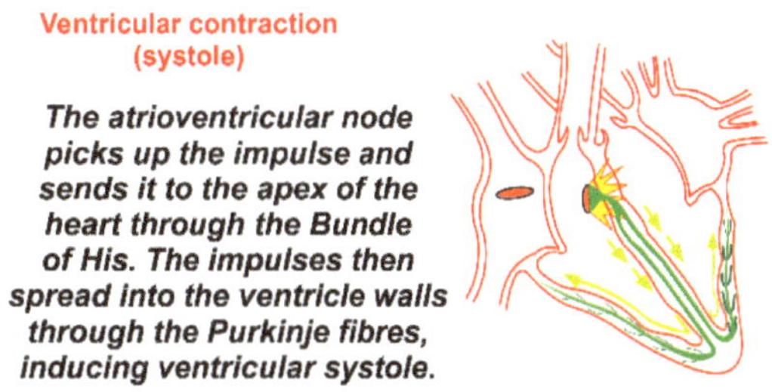

Figure 4.16: Ventricular Systole

The lack of conducting tissue between SAN and AVN causes a time delay. So that ventricular systole begins only after completion of atrial systole.

Edexcel IAL from 2018 Specification 1.9: (i) understand the role of haemoglobin in the transport of oxygen and carbon dioxide
(ii) understand the oxygen dissociation curve of haemoglobin, the Bohr effect and the significance of the oxygen affinity of fetal haemoglobin compared with adult haemoglobin

Transport of oxygen

- Oxygen is carried from the lungs to the tissues in the form of oxyhaemoglobin.
- Haemoglobin combines with O_2 in the lungs, where the O_2 partial pressure is high, to form oxyhaemoglobin.
- Oxyhaemoglobin is carried to the tissues, where the O_2 partial pressure is low and the CO_2 partial pressure is high. At a low partial pressure of O_2, haemoglobin has a lower affinity for O_2.
- So, oxyhaemoglobin dissociates to form oxygen and haemoglobin. The oxygen then diffuses into the tissues and is used for respiration.

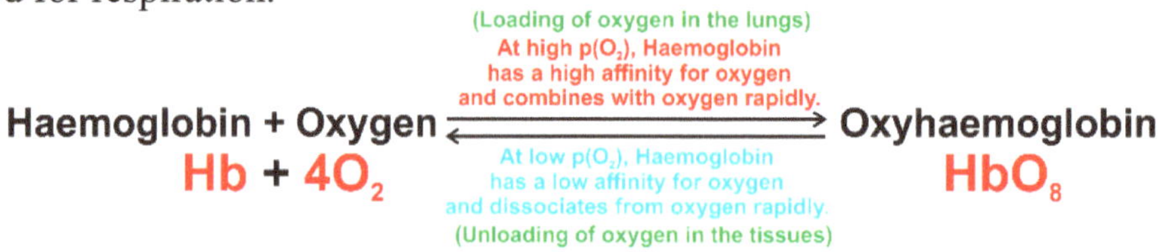

Transport of carbon dioxide

- About 5% of CO_2 dissolves in plasma and is transported from tissues to lungs as molecular CO_2.
- About 10% of CO_2 combines with the amino group of haemoglobin to form carbamino-haemoglobin, which dissociates in lungs to release CO_2.
- About 85% of CO_2 is carried in the form of hydrogen carbonates, dissolved in the plasma. The formation of bicarbonates or hydrogen carbonates is illustrated in figure 4.17.

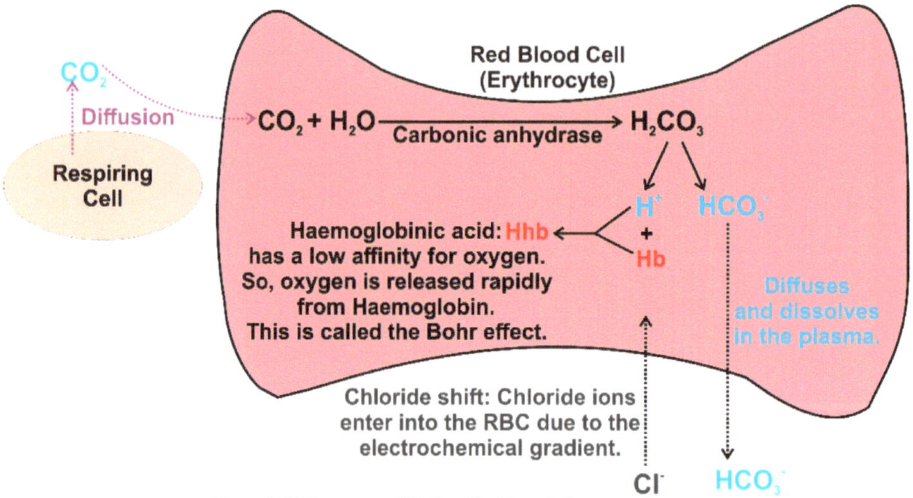

Figure 4.17: Transport of Carbon dioxide as hydrogen carbonates

Oxygen dissociation curve of haemoglobin

The oxygen dissociation curve shows the relationship between the concentration of oxygen or oxygen partial pressure and the percentage saturation of haemoglobin with oxygen, as shown in figure 4.18.

- Partial pressure is simply the pressure exerted by a gas in a mixture of gases.
- It is expressed as kPa (KiloPascals). It is dependent on the concentration of the gas in the mixture at a fixed temperature and volume.
- $p(O_2)$ represents the partial pressure of oxygen and $p(CO_2)$ represents the partial pressure of carbon dioxide.
- Loading tension is the $p(O_2)$ when Hb is 95% saturated with oxygen. 10.5 kPa is the loading tension in the graph in figure 4.18.
- Unloading tension is the $p(O_2)$ when Hb is 50% saturated. Each Haemoglobin molecule is combined with $2O_2$ molecules. 3.5 kPa is the unloading tension on the graph.

So, 45% of oxygen is released from the blood when it reaches the tissues.

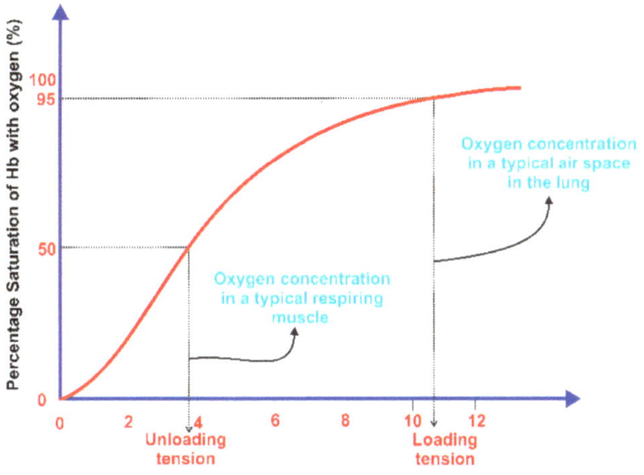

Figure 4.18: Oxygen dissociation curve

Each haemoglobin molecule consists of four polypeptide chains. Each polypeptide chain is linked to an iron containing haem group, as shown in figure 4.19. So, one haemoglobin molecule can combine with four oxygen molecules.

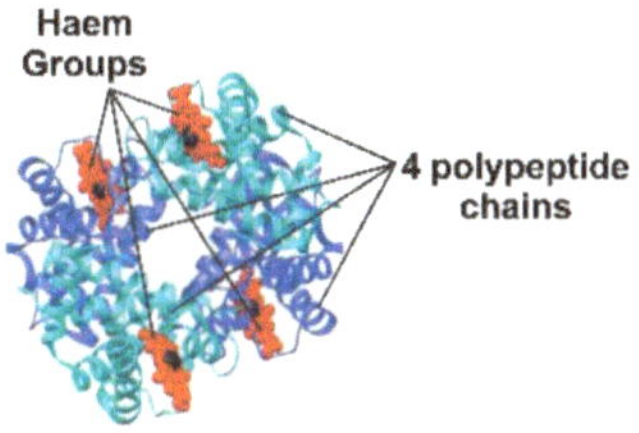

Figure 4.19: Haemoglobin Structure

The oxygen dissociation curve is sigmoid (S shaped). This is because **pure** haemoglobin has a low affinity for oxygen. So, the slope initially has a very low gradient. However, when haemoglobin combines with an oxygen molecule, the haemoglobin changes its shape (allosterism). This change in shape of the haemoglobin molecule makes it easier for the 2nd, 3rd and 4th O_2 molecules to combine with haemoglobin. This increasing ease with which haemoglobin combines with the remaining oxygen molecules is called **cooperativity**. The curve becomes steeper until all the haemoglobin molecules are completely saturated. The graph then flattens again, as shown in figure 4.20.

The steep part of the curve is very significant because it indicates that haemoglobin will release a lot of O_2 even for a small change in $p(O_2)$. This allows Hb to release O_2 to tissues more rapidly.

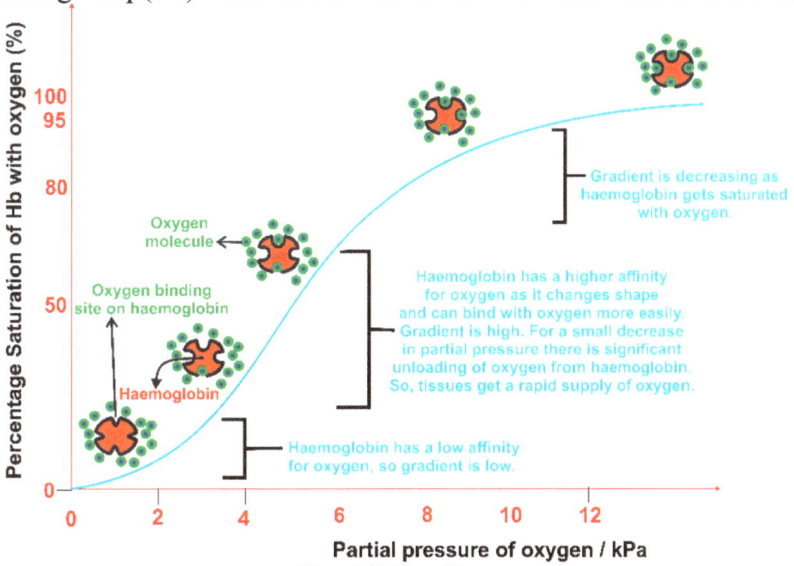

Figure 4.20: Sigmoid curve

Bohr Effect: this is the shifting of the O_2 dissociation curve to the right, due to an increase in the $p(CO_2)$. This increases the loading and unloading tension, so that haemoglobin has a lower affinity for O_2 (refer to transport of CO_2). The result is that haemoglobin gives up O_2 more readily to rapidly respiring tissues like the liver and muscles, where CO_2 concentration is high. The Bohr effect is also caused by a decrease in the pH of blood. The graphs in figures 4.21 and 4.22, shows the shifting of the curve as the blood moves between the lungs and the tissues.

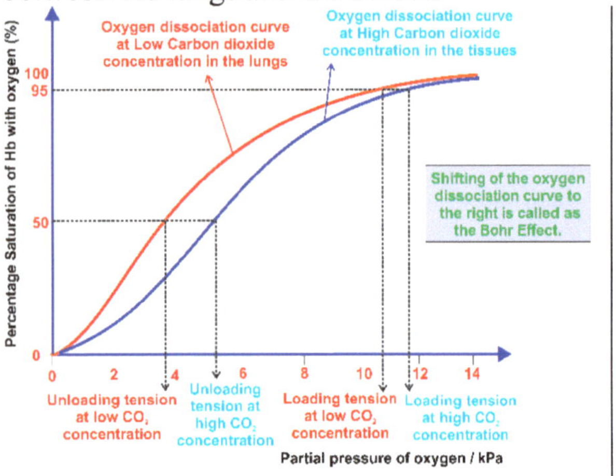

Figure 4.21: Bohr effect

The shifting of the oxygen dissociation curve to the right results in an increase in the loading and unloading tension of haemoglobin. The higher loading tension means that a higher oxygen partial pressure will be needed to saturate 95% of haemoglobin with oxygen. Likewise, a higher unloading tension means that a higher oxygen partial pressure will be needed to saturate 50% of haemoglobin with oxygen. This indicates that haemoglobin has a lower affinity for oxygen due to the Bohr effect.

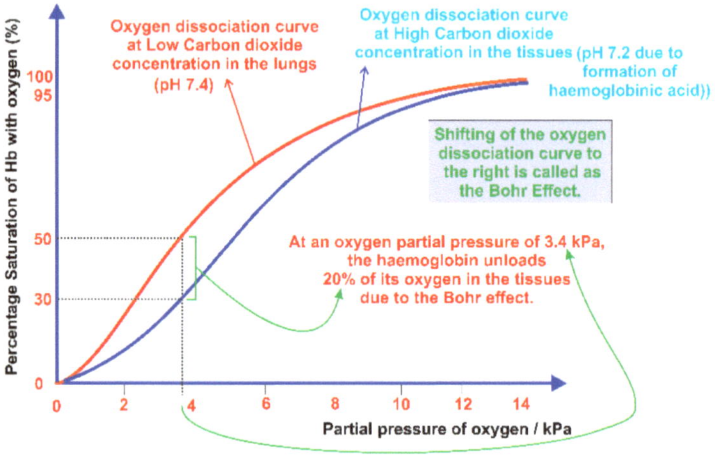

Figure 4.22: Bohr effect

The shifting of the oxygen dissociation curve to the right results in the haemoglobin being less saturated with oxygen at any given partial pressure of oxygen. At an oxygen partial pressure of 3.5 kPa, in Figure 4.21, the percentage saturation of haemoglobin decreases by 20%. This means that haemoglobin will release 20% of its oxygen to the tissues when the Carbon dioxide concentration increases.

The decrease in affinity for oxygen is due to the formation of Haemoglobinic acid (pH 7.2) is shown in figure 4.17.

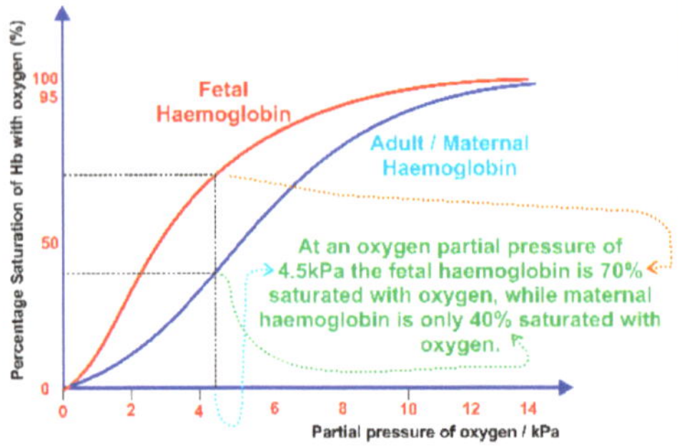

Figure 4.23: Oxygen dissociation curve for fetal haemoglobin

Fetal hemoglobin has a greater affinity for oxygen, compared to adult hemoglobin. Since the fetal haemoglobin has a higher affinity for oxygen, the curve will lie to the left of the curve for maternal haemoglobin. Fetal haemoglobin will have a lower loading and unloading tension than maternal haemoglobin. This allows fetal hemoglobin to take up O_2 more readily than maternal hemoglobin at any given $p(O_2)$.

For example, the graph in figure 4.23 shows that fetal haemoglobin is 30% more saturated with oxygen than the maternal haemoglobin, at a $p(O_2)$ of 4.5kPa. This ensures that the fetus can get a continuous supply of O_2 from maternal blood in the placenta.

CHAPTER FIVE
CARDIOVASCULAR DISEASES

Edexcel IAL from 2018 Specification 1.10: understand the course of events that leads to atherosclerosis (endothelial dysfunction, inflammatory response, plaque formation, raised blood pressure)

How Atherosclerosis Develops

Endothelial Dysfunction: The wall of an artery is composed of several layers. The lining or inner layer (endothelium) is usually smooth and unbroken. Atherosclerosis is caused by repeated injury to the endothelial layer in the walls of arteries. The **endothelium is injured or diseased**. Many factors contribute to this injury, including **high blood pressure, tobacco smoke, diabetes** and **high levels of cholesterol** in the blood.

Inflammatory response: certain white blood cells called monocytes and T cells are activated and move out of the bloodstream and through the lining of an artery into the artery's wall, this is called as the inflammatory response. Inside the lining, they are transformed into **foam cells**, which are cells that collect fatty materials, mainly cholesterol. In time, smooth muscle cells move from the middle layer into the lining of the artery's wall and multiply there.

Plaque formation: Connective and elastic tissue materials also accumulate there, as may cell debris, cholesterol crystals, and calcium. This accumulation of fat-laden cells, smooth muscle cells, and other materials forms a patchy deposit called an **atheroma** or **atherosclerotic plaque.**

Raised blood pressure: As they grow, some plaques thicken the artery's wall and bulge into the channel of the artery. This hardened area means that part of the artery wall **hardens**, so it is **less elastic** than it should be. This is **atherosclerosis.**
The lumen of the artery becomes much smaller and less elastic as a result of the plaque. This increases the blood pressure, making it harder for the heart to pump blood around the body. The raised blood pressure makes damage more likely in the endothelial lining in other areas, and more plaques will form. More plaques will make the blood pressure even higher. This is an example of **positive feedback, as shown in figure 5.1.**

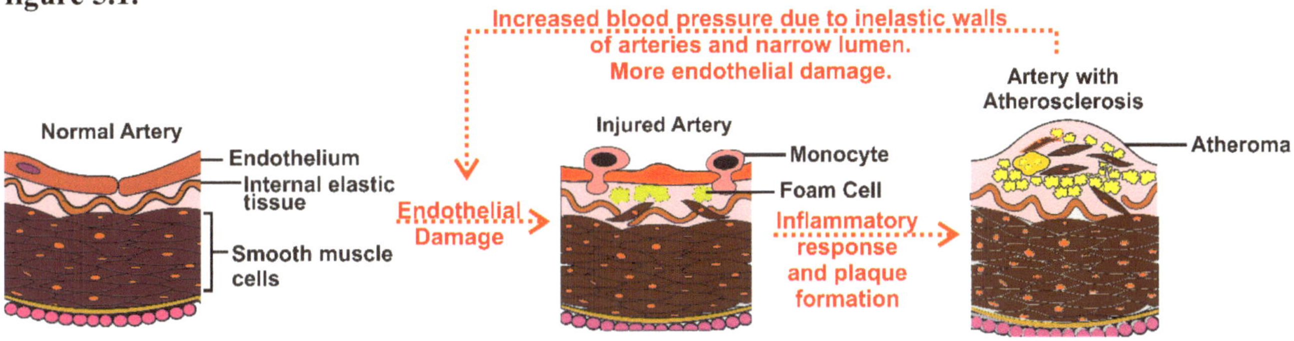

Figure 5.1: Atheroma formation

These plaques may narrow or block an artery, reducing or stopping blood flow. Other plaques do not block the artery very much but may split open, triggering a blood clot that suddenly blocks the artery. Plaques may be scattered throughout medium-sized and large arteries, but they usually form where the arteries branch due to turbulence of blood flow at these regions.

Edexcel IAL from 2018 Specification 1.11: **understand the blood clotting process (thromboplastin release, conversion of prothrombin to thrombin and fibrinogen to fibrin) and its role in cardiovascular disease (CVD)**

Blood clotting – Thrombosis

Platelets are fragments of cells broken off from large cells (called megakaryocytes) in the bone marrow. They play an important role in blood clotting. The process of blood clotting is illustrated in figure 5.2.

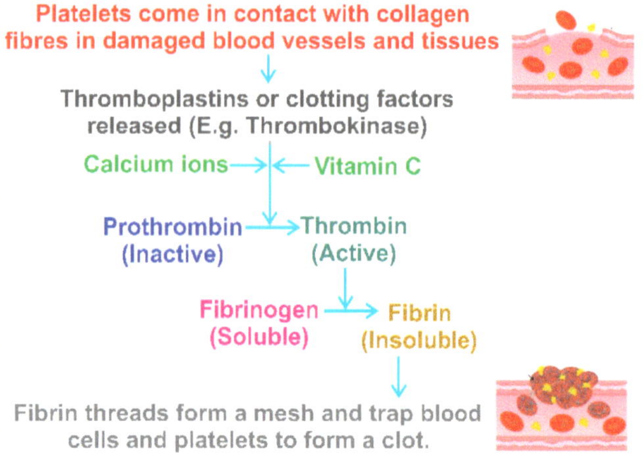

Figure 5.2: Blood clotting

Platelets stick to a site of damaged tissue. Thromboplastin is released from this area, which results in a plasma protein called prothrombin being converted to thrombin. Thrombin acts as an enzyme and catalyses the conversion of the soluble plasma protein fibrinogen into long, insoluble strands of the protein fibrin These strands form a mesh that traps red blood cells to form a blood clot. Blood clotting minimizes blood loss following injury. The blood coagulates to form a solid plug (clot) made or cells trapped in a fibrous network.

What causes blood to clot inside blood vessels?

Blood **does not normally clot in intact blood vessels** because of the action of several anticoagulants such as heparin circulating in the bloodstream. Intact endothelium (the inner lining of blood vessels) also produces molecules which inhibit clotting. Blood clots quickly when exposed to air because of the absence of an endothelium and a lack of anticoagulants. A **damaged endothelium** or presence of **certain chemicals** (from cigarette smoke) in the blood causes the blood to clot (Thrombosis).

Why does blood clot in arteries more often than in veins?

Blood clotting is initiated when the platelets come in contact with collagen in the walls of damaged blood vessels. The platelets get activated and their shape changes. The blood in arteries flows with a high pulsating pressure than in veins. The high pulsating pressure in arteries leads to more damage in the endothelium, increasing the chances atherosclerosis and internal clots.

The **clot**

- **prevents further blood loss,**
- **reduces the risk of pathogens (harmful micro-organisms) entering the body, and**
- **provides a framework for the repair of damaged tissue.**

Blood clotting is an essential process, for the reasons stated above, however if the blood clots inside the arteries then it would increase the risk of suffering from heart attacks and strokes. For example, some people carry the **platelet antigen gene** which causes platelets to be overly active in forming blood clots. These clots could block the arteries and increase the risk of cardiovascular diseases. Other factors, like high blood pressure can damage the endothelium and increase the risk of clotting in arteries, increasing the risk of heart attacks and strokes.

Role of blood clotting and atherosclerosis in cardiovascular disease (CVD)

Cardiovascular diseases (CVD) are diseases associated with the coronary circulation or diseases of the major arteries, like arteries in the brain or limbs. The sequence of events is shown in figure 5.3 shows the role of atheroma and thrombosis in a heart attack.

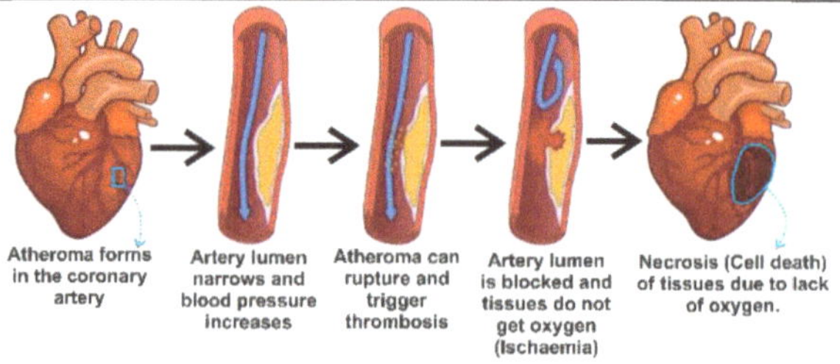

Figure 5.3: Role of blood clots in Coronary Heart Diseases

Cardiovascular diseases (CVDs) is a term used to describe diseases that involve the heart or blood vessels. The chart in figure 5.4 shows the consequences of atherosclerosis and thrombosis.

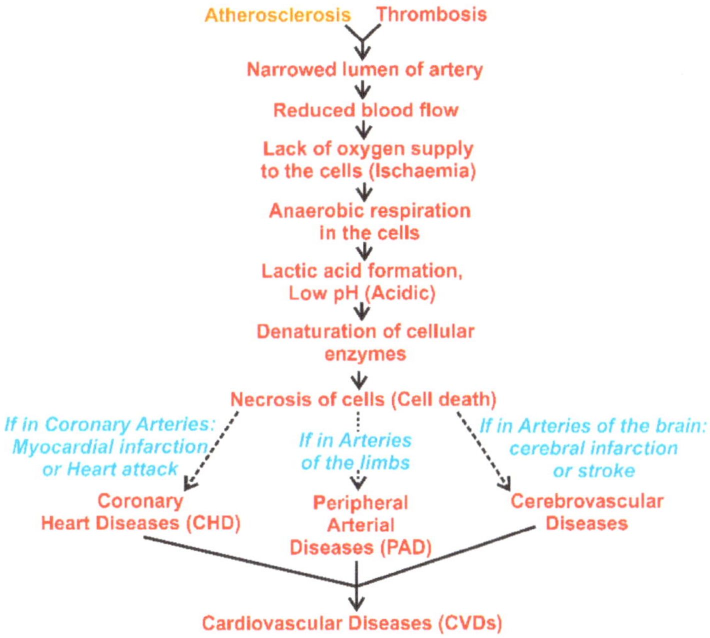

Figure 5.4: Consequences of atherosclerosis and thrombosis

Coronary heart disease (CHD)

If blood clotting and atherosclerosis occurs inside a coronary blood vessel, then the vessel gets blocked and results in lack of blood flow to the cardiac muscles. The clotting of blood in the coronary artery is called as **coronary thrombosis**. This will lead to lack of oxygen and glucose in the cardiac muscle tissue. This is referred to as **ischaemia**. Ischaemia means local anaemia in a given body part. The anaerobic conditions will cause the muscles to stop working and damaged cells die. Lactic acid builds up in the cardiac muscles and results in pain called as **angina pectoris**. This may also lead to myocardial infarction. **Myocardial infarction is** a heart attack. The main change is necrosis (death) of myocardial tissue.

Cerebrovascular diseases (Lead to strokes): If blood clotting occurs inside an artery supplying blood to the brain, then it can result in stroke. The stroke caused by the clotting of blood inside the artery is called as an ischaemic stroke. Sometimes the arteries may rupture in the brain, leading to a haemorrhagic stroke, as shown in figure 5.5.

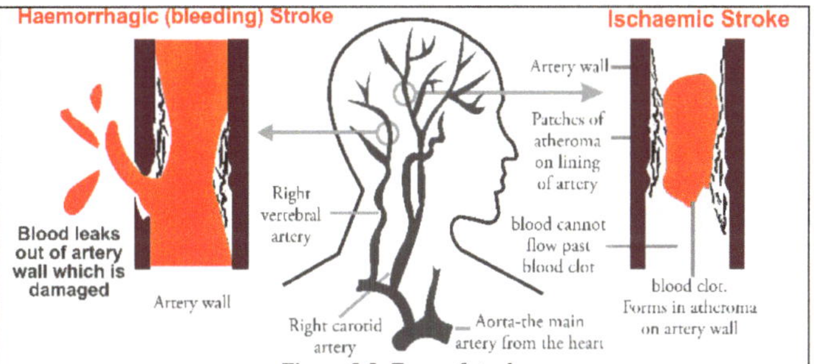

Figure 5.5: Types of stroke

The symptoms of strokes vary, depending on how much of the brain is affected. Very often the blood is cut off from one part or one side of the brain only. Symptoms include dizziness, confusion, slurred speech, blurred vision or loss of part of the vision (usually just in one eye) and numbness. In more severe strokes there can be paralysis, usually just down one side of the body.

Peripheral arterial disease (PAD): This occurs when plaque builds up in the major arteries that supply oxygen-rich blood to the legs, arms, and pelvis. When blood flow to these parts of your body is reduced or blocked, it can lead to numbness, pain, and sometimes dangerous infections.

Self-assessment (Chapter 5)

Step 1: Visit www.teachingtube.com

Step 2: Sign up for the free course "Edexcel AS and AL Bio assessments"

Step 3: Attempt the chapter Quiz and get instant results.

Step 4: Download the assessment worksheet and marking scheme too.

Chapter Six
Risk factors

Edexcel IAL from 2018 Specification 1.12: know how factors such as genetics, diet, age, gender, high blood pressure, smoking and inactivity increase the risk of cardiovascular disease (CVD)

Measuring or estimating risk

A risk factor is anything that affects or raises the **chance** of harm. **Chance** or **probability** can be measured in terms of percentage or ratio.

Risk describes the probability that an event will happen. Probability means the chance or likelihood of the event, calculated mathematically.

In 2005, 335969 people died due to Coronary Heart Diseases (CHD). The total UK population at that time was 60 209 408, so we can calculate the average risk in a year of someone in the UK dying from Coronary Heart Diseases (CHD) as:

$$\text{Risk} = (\text{Number of events} / \text{total entities}) \times 100$$
$$\text{Risk} = (335969 / 60\ 209\ 408) \times 100 = 0.55\%$$

Cholesterol, LDL, HDL and Triglycerides

Cholesterol

*Cholesterol is a fat-like lipid that is found in all body cells. Cholesterol is an essential nutrient required structurally for cell membranes and myelin sheaths. Cholesterol in the body comes from two sources. A small amount is eaten in certain foods, but most is produced by our Liver, and we need some in our bodies to stay healthy. The liver produces cholesterol when **saturated fats** are eaten. If the diet contains a lot of saturated fats, then the liver produces more cholesterol than the body requires. Since, Cholesterol is insoluble in the blood it travels to cells through the bloodstream in special carriers called **lipoproteins**. Two of the most important lipoproteins are **low-density lipoprotein** (LDL – carries bad cholesterol) and **high-density lipoprotein** (HDL – carries good cholesterol).*

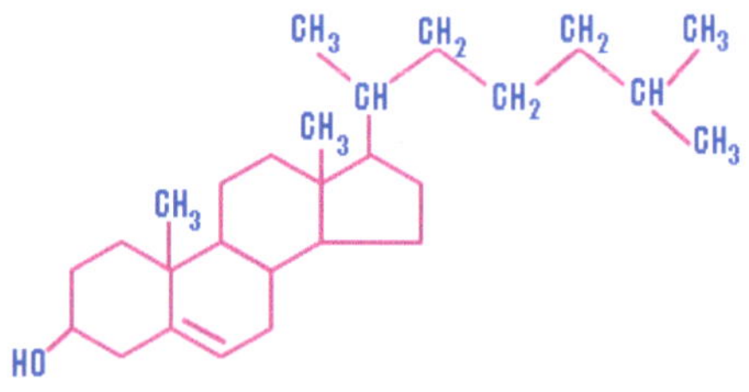

Figure 6.1: Structure of cholesterol (Not required for exams):

*A lipoprotein is made of a single layer of phospholipids with their hydrophilic heads facing the surface of the particle. Apolipoproteins are embedded into the phospholipid layer. Cholesterol and triglycerides are carried in the core of the particle, as shown in figure 6.2. Lipoproteins differ in their content of proteins and lipids. **The higher the ratio of protein to lipid content the higher the density.***

Low-density Lipoprotein

*LDL particles deliver cholesterol to the body's cells. LDL cholesterol is often called "bad cholesterol" because high levels are thought to cause fatty plaque on artery walls, which starts the process of **atherosclerosis**. The LDLs circulate in the blood stream and bind to receptors on the cell surface membrane before being taken up by the cells.*

High-density Lipoprotein

HDL particles carry cholesterol from the body's cells back to the liver, where it can be removed from the body. HDL is known as "good cholesterol" because high levels are thought to lower the risk of heart disease.

Triglycerides

Triglycerides are fats that provide energy for the muscles. Like cholesterol, they are delivered to the body's cells by lipoproteins in the blood. Eating foods with a lot of saturated fat or carbohydrates will raise triglyceride levels. Elevated levels are thought to lead to a greater risk of heart disease, but scientists do not agree that high triglycerides alone are a risk factor for heart disease.

Key:
ApoA, ApoB, ApoC, ApoE (apolipoproteins – protein components);

T (triacylglycerol);

C (cholesterol)

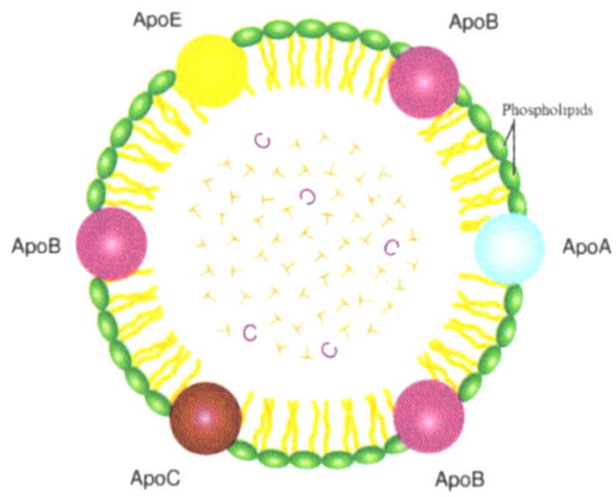

Figure 6.2: Structure of lipoprotein

Risk factors for cardiovascular diseases

Any factor which increases the chances of suffering from cardiovascular diseases is called as a risk factor.

1. Genetic factors

a. Familial hypercholesterolemia (FH) – increasing risk in children and teenagers

In some cases, high cholesterol runs in families. This is called **familial hypercholesterolemia (FH)**. FH is characterized by high cholesterol levels, specifically very high low-density lipoprotein (LDL, "bad cholesterol") levels, in the blood and early cardiovascular disease (CVD).

FH is caused due to a defective *LDLR* gene that encodes the LDL **Receptor** protein, which normally removes LDL from the circulation, as shown in fig.6.3.

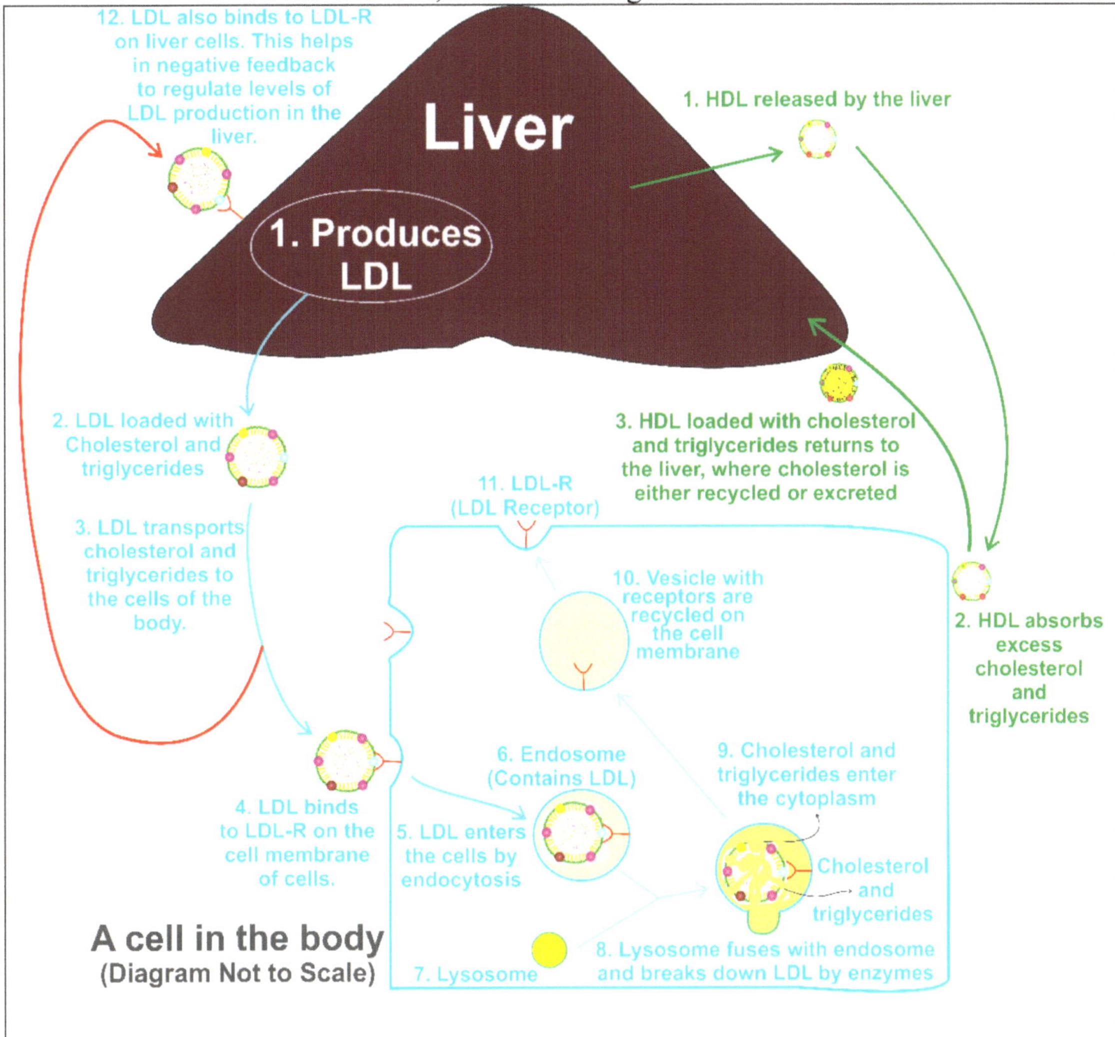

Figure 6.3: Transport of cholesterol and triglycerides by LDL and HDL

If the LDL-R gene is mutated, then it will code for a defective receptor. The defective LDL-Receptor cannot bind to LDL or cannot remove LDL from the blood. So, the blood LDL levels remain high and increase the risk of CVDs.

There are five major classes of FH due to LDL-R mutations:

Class I: LDL receptor (LDL-R) is not synthesised at all

Class II: LDL-R is synthesised but does not bind to the cell membrane

Class III: LDL-R does not properly bind LDL on the cell surface

Class IV: LDL-R bound to LDL does not enter cells by endocytosis

Class V: the LDL-R is not recycled back to the cell surface.

b. APOE (4) and APOE (2) Genes

Another cluster of genes that is associated with coronary heart disease is the apolipoprotein gene.

Apolipoproteins are the protein component of lipoproteins. They are formed in the liver and intestine. The main role is in stabilizing the structure of lipoproteins, shown in fig.6.2, and recognizing receptors involved in lipoprotein uptake on the plasma membrane of most cells in the body.

There are a few different types of Apolipoproteins.

Type of Apolipoprotein	Function and consequences of malfunction
Apolipoprotein A (APOA)	the main protein in HDL, which helps in the removal of cholesterol to the liver for excretion. Mutations in the apoA gene are linked to low HDL levels and reduced removal of cholesterol from the blood, leading to increased risk of coronary heart disease.
Apolipoprotein B (APOB)	the main protein in LDL, the molecule which transfers cholesterol from the blood to cells. Mutations of the apoB gene result in higher levels of LDL in the blood and a higher risk of CVD.
Apolipoprotein E (APOE)	a major component of HDLs and very low-density lipoproteins (VLDLs), which are also involved in removal of excess cholesterol from the blood to the liver. The apoE gene has three common alleles, producing three forms of the protein, E2, E3 and E4. APOE4 slows removal of cholesterol from the blood and therefore having the E4 allele may increase the risk of coronary heart disease.

c. The platelet antigen gene linked to sudden death in athletes.

The platelet antigen gene is a form of a gene which greatly increases the chances of a heart attack. This form of the gene causes platelets to be overly active in forming blood clots and may cause cholesterol to bind to endothelial cells lining blood vessels. Increased deposition of cholesterol in the walls of coronary arteries will lead to more rapid development of atheroma and narrowing of the blood vessel. As the blood slows, the sticky platelets are more likely to form a clot. Any damage to the vessel walls will result in rapid formation of a clot, leading to sudden death.

2. Diet

The following recommendations are made by the **British Nutrition Foundation** in order to reduce the risk of cardiovascular diseases.

Eat a healthy, varied diet

- Use vegetable oil that is high in unsaturated fat in cooking, but only in small amounts, *e.g.* olive oil or rapeseed oil.
- Choose lean meat, poultry, beans and alternatives instead of fatty meat or meat products.
- Eat less fat and fatty foods

Explanation

Polyunsaturated fatty acids (Vegetable oils) 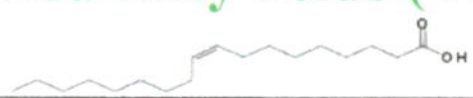	Saturated fatty acids (Animal fats) 
Vegetable oil contains high percentage of polyunsaturated fatty acids, which form HDLs.	Animal fat contains high percentage of saturated fatty acids, which form LDLs.
Hence, vegetable oils (HDLs) reduce blood cholesterol and reduce the chance of forming atherosclerosis.	Hence, animal fats (LDLs) contribute to the build-up of atheroma or plaque, causing atherosclerosis.

Lipid rich food also contains a lot of energy. All this energy may not be used by the body and the excess will be stored in adipose tissue. This increases the risk from **obesity related** disorders.

- Eat fish at least twice a week, of which one portion should be oily fish. 

Explanation

Fish oil contains *omega-3 fatty acids*, a group of polyunsaturated fatty acids with their first bond between the third and fourth carbon atoms. These fatty acids are essential for cell functioning and have been linked to a reduction in atherosclerosis and joint inflammation.

• Choose low-salt products and use less salt in cooking. 

Explanation

Excess salt can cause an increase in the retention of water in the blood by the kidneys. This can increase the volume of blood and blood pressure, increasing the risk of Cardiovascular diseases.

• If you drink alcohol, do so moderately. The guidelines for safe intakes are no more than 3-4 units per day for men and 2-3 units per day for women (1 unit = 8g).

Explanation

Heavy drinking raises blood pressure, contributes to obesity and can cause irregular heartbeat. It also damages liver cells.

Moderate consumption of alcohol has a beneficial effect on health. It is thought to increase the HDL levels in blood. Red wine contains antioxidants, which reduce the risk of cardiovascular diseases.

3. Age

Prevalence of CVD increases with age, for example, data from 2011 suggest that about 35% of men over 75 years were affected by CVD compared with about 15% of 45-64 year olds.

The following changes take place as the person ages:

Narrowing of arteries and reduced elasticity cause thickening of the left ventricle

Researchers have noted that the wall of the left ventricle of the heart becomes thicker with age, as shown in figure 6.4. This thickening allows the heart to pump stronger. As our blood vessels age, they become narrower, causing blood pressure to increase. The heart compensates for this by becoming stronger and pumping with more force. The thick wall of the left ventricle occupies more space and the stroke volume will increase. The heartbeat rate will therefore have to increase to ensure the required cardiac output. The increased blood pressure and increased heartbeat rate increase the risk of myocardial infarction.

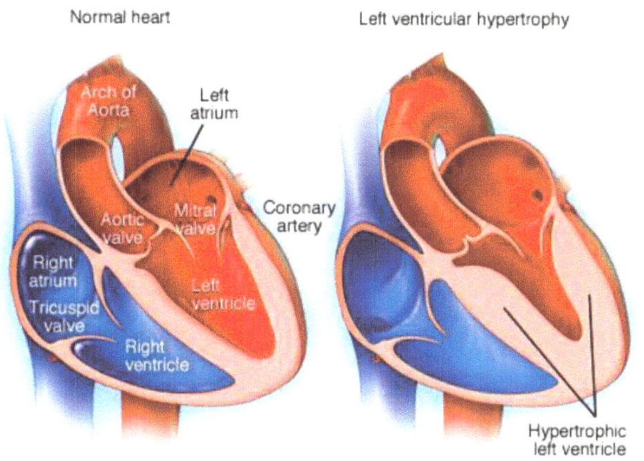

Figure 6.4: Thickening of the left ventricle

Exercise Capacity Shrinks

As the heart ages, it becomes less able to respond rapidly to chemical messages from the brain. Researchers do not know exactly why the heart does not respond as fast to messages to speed up and adjust to increased activity. The result is the body cannot exercise as long or as intensely as before. This shows up as shortness of breath, a sign that oxygen-rich blood is not moving fast enough through the body because the lungs are trying to breath in more oxygen.

4. Gender

Before menopause, a woman's own oestrogen helps protect her from heart disease by increasing HDL (good) cholesterol and decreasing LDL (bad) cholesterol. After menopause, women have higher concentrations of total cholesterol than men do. But this alone doesn't explain the sudden rise in heart disease risk after menopause. Elevated triglycerides are an especially powerful contributor to cardiovascular risk in women. Low HDL and high triglycerides appear to be the major factor that increase the risk of death from heart disease in women over age 65.

5. High blood pressure

Blood pressure is usually expressed in terms of the systolic pressure over diastolic pressure, for example 120/80. If the pressure remains consistently higher than normal for an extended period, then it is referred to as hypertension or high blood pressure. **High blood pressure is the No. 1 modifiable risk factor for stroke.** It also contributes to heart attacks, heart failure, kidney failure and atherosclerosis. The relationship of blood pressure levels to the risk of cardiovascular disease is continuous, consistent and independent of other risk factors. That means the higher your blood pressure, the greater your risk of heart attack, heart failure, stroke and kidney disease. The factors affecting blood pressure are shown in figure 6.5.

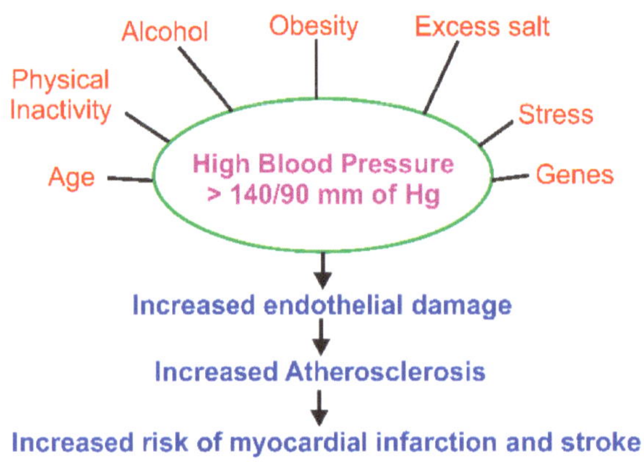

Figure 6.5: High blood pressure

6. Cigarette smoking

The effects of smoking and the causal links are shown in figure 6.6.

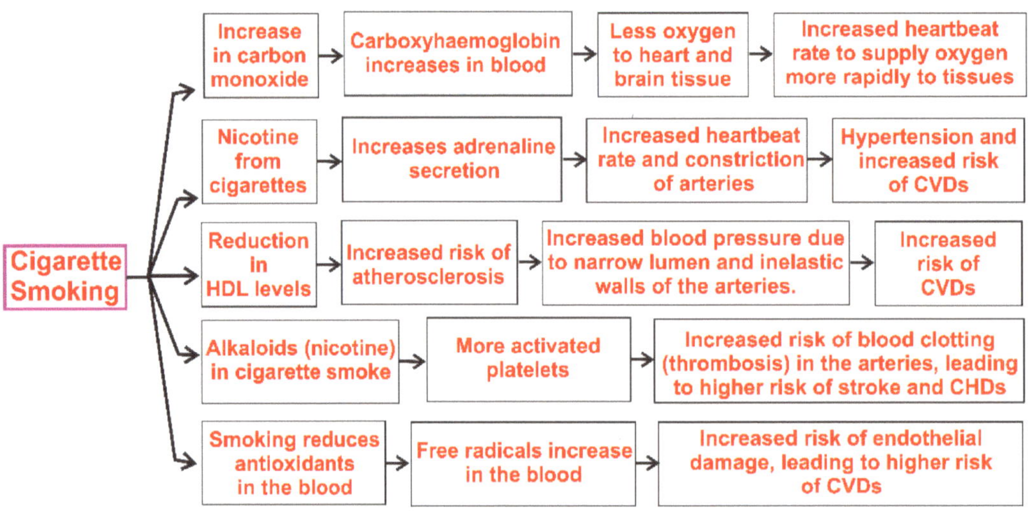

Figure 6.6: Effects of smoking

7. Inactivity - Lack of exercise

Lack of exercise is accepted as a major risk factor of CHD. As little as, 8 to 16 km of walking or jogging each week can offer some protection against the disease. Aerobic exercise strengthens the heart, enabling it to contract more forcefully so that, for any given workload, the stroke volume increases and the heart rate decreases. Exercise does not guarantee protection from CHD - it just reduces the risk. Exercise also increases HDL levels. Physical activity burns fat, reducing the risk of obesity and plaque formation in blood vessels.

Multifactorial disease

A multifactorial disease is one whose development is affected by a range of factors, for example heredity, physical environment, social environment and lifestyle choices all contribute to the risk of developing CVD. Each factor increases the risk of suffering from CVD, but caution must be taken in making conclusions about the cause of CVD.

Edexcel IAL from 2018 Specification 1.13: **understand the link between dietary antioxidants and the risk of cardiovascular disease (CVD)**

Free radicals are charged particles produced by metabolism in the body. They attach to cells and cause cellular damage, as shown in figure 6.7.

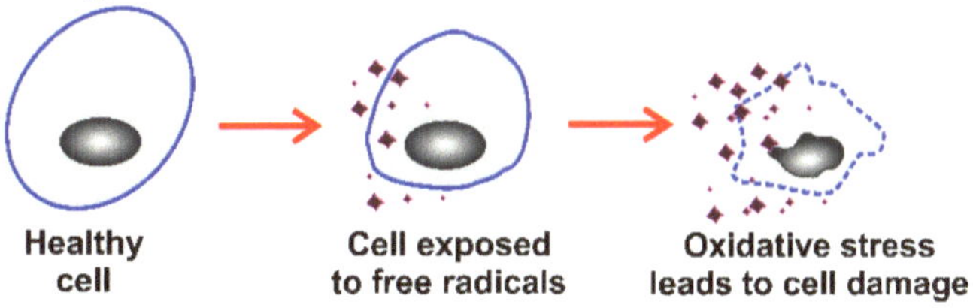

Figure 6.7: Effect of free radicals on cells

Fruits and vegetables contain antioxidants like Vitamin C and beta carotene. The antioxidants neutralise free radicals and prevent cellular damage. Antioxidants can neutralise these radicals and reduce the harm caused to the cells (especially myocardial cells and endothelium), as shown in figure 6.8.

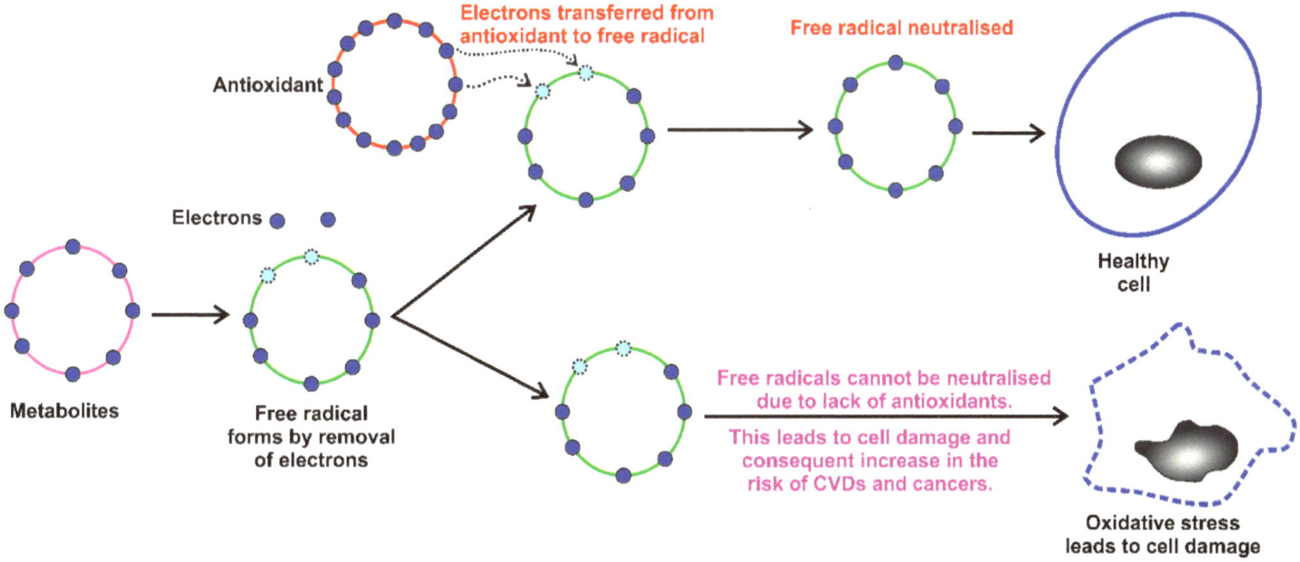

Figure 6.8: Antioxidants reduce the risk of oxidative stress

Step one: Preparation of Vitamin C solution

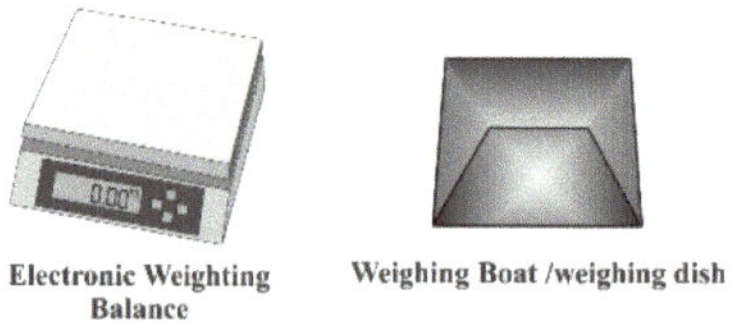

Figure 6.9 – Weighing the Vitamin C

☆ Weigh 10mg of pure vitamin C powder using a weighing boat and an electronic balance.

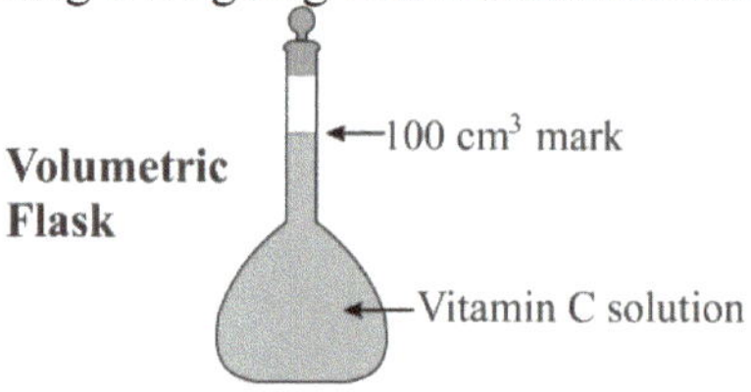

Figure 6.10 – Preparing the vitamin C solution

☆ Add distilled water into the vitamin C until the volume of the solution reaches 100cm^3.
☆ The concentration of the vitamin C solution can be calculated as follows.

Concentration (C_A) = mass of vitamin C powder / volume of solution

10 / 100 = 0.1 mg cm^{-3}

Step two: Preparation of DCPIP Solution

☆ Dissolve 10mg of DCPIP into distilled water till the volume reaches 100 cm^3.
☆ The resulting solution has a concentration of 0.1mg cm^{-3}

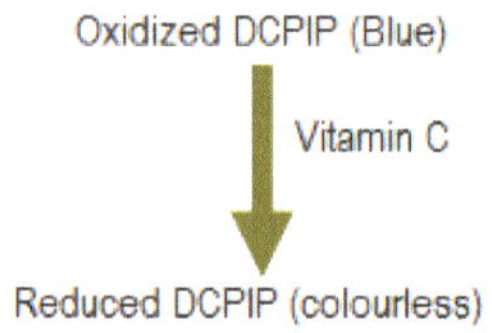

Note: Use the same procedure as step one to prepare the solution.

Step three: Estimate the volume of vitamin C solution needed to decolorize 2cm^3 of the DCPIP

A graduated pipette may be used to drop the Vitamin C solution into the DCPIP.

Figure 6.11 – A graduated pipette

This volume is referred to as V_A

Step four: Extraction of fruit juice

Follow the steps in the sequence shown below:

Peel the fruit → Cut and weigh specific mass of fruit → Blend in a blender to form pulp → Filter and collect the Filtrate (juice).

Step five: Determine volume of fruit juice needed to decolorize 2cm^3 of the DCPIP (V_B)

Use the same method as in step three but use fruit juice instead of vitamin C solution.

Step six: Use the equation below to estimate the concentration of vitamin C in the fruit juices.

$$CONC_B = (VOL_A \times CONC_A) / VOL_B$$

VOL_A = Volume of vitamin C solution in cm^3
$CONC_A$ = Concentration of vitamin C solution in mg cm^{-3}
VOL_B = Volume of fruit juice in cm^3
$CONC_B$ = Concentration of vitamin C in fruit juice in mg cm^{-3}

Self-assessment (Chapter 6)

Step 1: Visit www.teachingtube.com
Step 2: Sign up for the free course "Edexcel AS and AL Bio assessments"
Step 3: Attempt the chapter Quiz and get instant results.
Step 4: Download the assessment worksheet and marking scheme too.

CHAPTER SEVEN

CORRELATION, CAUSATION AND STUDY DESIGN

Edexcel IAL from 2018 Specification 1.15: be able to analyse and interpret quantitative data on illness and mortality rates to determine health risks, including distinguishing between correlation and causation and recognising conflicting evidence

Correlation and causation

Correlation is when a change in one variable is **reflected** by a significant change in another variable.

Causation is when a change in one variable is **responsible** for a change in another variable.

Any scientific study requires the collection and interpretation of data. It is often necessary to establish whether there is a significant relationship between two variables. This relationship can be statistically determined using correlation coefficients. The correlation coefficient can be interpreted to determine whether there is a positive correlation, negative correlation or no correlation at all between the two variables.

A positive correlation means that variable 2 will increase as variable 1 increases, as shown in figure 7.1.

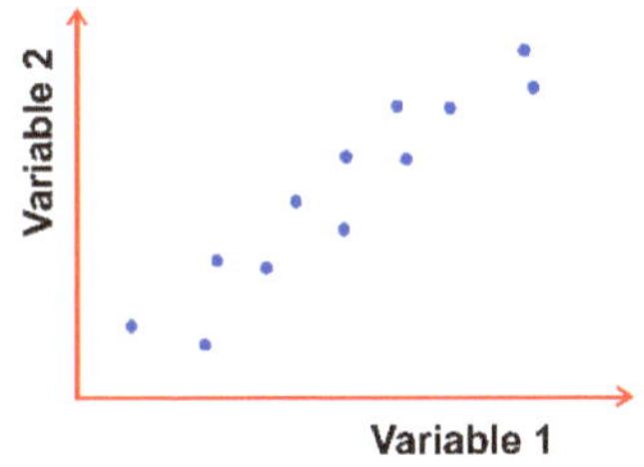

Figure 7.1: Positive correlation

For example, the relationship between the amount of annual rainfall and the yield of crops may show a positive correlation.

However, it would be inappropriate to conclude that an increase in annual rainfall causes the yield of crops to increase. This is because other factors, like temperature, mineral content of soil, soil type, light intensity and a range of other factors could also influence the yield of crops. However, if these factors are controlled then it may be more indicative that the volume of annual rainfall causes the crop yield to increase.

A negative correlation implies that variable 2 will decrease as variable 1 increases, as shown in figure 7.2.

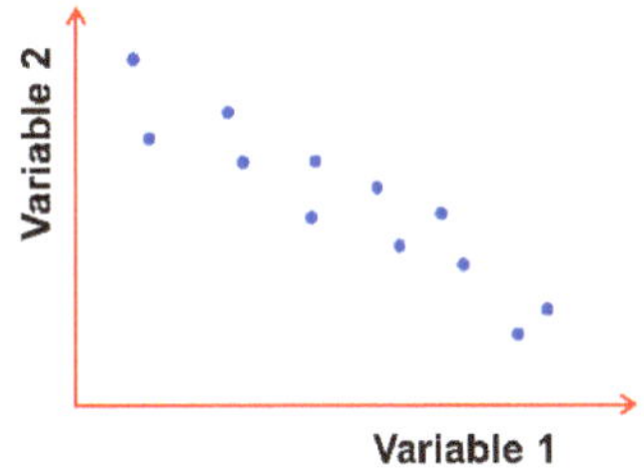

Figure 7.2: Negative correlation

For example, the relationship between the number of cigarettes smoked per day and the lung capacity may show a negative correlation.

However, it would be inappropriate to conclude that an increase in the number of cigarettes smoked per day causes the lung capacity to decrease. This is because other factors, like exercise, age, gender, genetics and a range of other factors could also influence the lung capacity. However, if these factors are controlled then it may be more indicative that an increase in smoking frequency causes crop yield to decrease.

A correlation does not imply causation. A causal link can be established only when we can provide a clear explanation of how one factor may be responsible for a change in another factor. For example, it is clearly understood that mutant LDL-R genes can result in defective LDL-R on the cell surfaces. This in turn prevents the absorption of LDL into the cells and the blood LDL levels remain high. So, there is a clear causal link between mutant LDL-R genes and high blood LDL cholesterol levels.

A correlation coefficient will simply confirm whether the two variables are correlated or not. It does not imply any causation. Further investigation needs to be done to identify the causative agent.

Data analysis and conflicting evidence – an example

The table below shows the relationship between blood cholesterol levels and death from coronary heart disease in men in the UK.

Serum cholesterol / mmol dm^{-3}	Frequency in population	Deaths from CHD per 1000 men each year
4.1 – 4.9	22	2
4.9 – 5.7	26	5
5.7 – 6.5	30	8
6.5 – 7.3	12	12
7.3 – 8.2	9	18
8.2 – 8.8	4	25
8.8 – 9.6	4	30

The data clearly shows that as the **serum cholesterol** levels increase, the **death from CHD** also increases. There is **a correlation between the two factors**, however **high serum cholesterol may not be the cause** for the increased incidence of CHD.
In recent years we have learned that it is not the cholesterol in the diet that causes CHDs, but it is the relative levels of HDL and LDL in the blood which is the **cause** of plaque formation and CHDs.

Edexcel IAL from 2018 Specification 1.16: be able to evaluate the design of studies used to determine health risk factors, including sample selection and sample size used to collect data that is both valid and reliable

Features of a good study

To establish a correlation and identify the causative agent of various diseases data collection and interpretation is vital. The design of the study and methods of data collection will influence the validity and reliability of the investigation. The following criteria can be used as a guide to obtain data that is both valid and reliable.

1. Clear aim

A well-designed study should include a clearly stated hypothesis or aim. The hypothesis should be testable, and the results must be analysed in a valid and reliable way, using statistics.

2. Representative sample

a. Large sample size

The sample size must be large enough to provide statistical significance. It should usually represent a large proportion of the population.

b. Randomised selection

The selection of individuals must be done by randomization to reduce bias (the inclination or tendency to favour one point of view). In randomization, the patients are allocated by chance to the different groups. This procedure is performed with random numbers from a calculator or by picking numbers from a hat. Randomization ensures that the patients will be allocated to the different groups in a balanced manner and that possible confounding factors — such as risk factors and genetic variation — will be distributed by chance between the groups. Randomization is intended to maximize homogeneity between the groups and prevent, for example, a specific therapy being reserved for patients with a particularly favourable condition (such as young patients in good physical condition).

c. Control group

There must also be a control group to compare the results and draw valid conclusions.

3. Duration

The study design should be such that the study can be completed in a feasible time span. If the study is too long, then the proportion of individuals who drop out of the study may be too high. This could give rise to complications while analysing the results and reduce the validity of the study.

4. Blinding

Blinding is another suitable method to avoid bias. Blinding means that neither the patient nor the investigator knows the treatment being administered. In single blinding, the patient is unaware which treatment he is receiving, while, with double blinding, neither the patient nor the investigator knows which treatment is planned.

Blinding the patient and investigator excludes possible subjective (even subconscious) influences on the evaluation of a specific therapy (e.g. drug administration versus placebo). Thus, double blinding ensures that the patient or therapy groups are both handled and observed in the same manner. The highest possible degree of blinding should always be selected. The control group is usually given a 'fake drug' or 'sugar pill' called as the **placebo**. The placebo is a 'make-believe' treatment. The case group is given the actual treatment.

5. Valid and reliable results

A reliable method for collection and analysis of data will produce similar results when used by different people at different times for repeated measurements. **All variables that can affect the results must be controlled or adjusted**.

The Munster study on cardiovascular diseases had the following features.

Aim

To identify the causative and contributory agents of Cardio-vascular diseases.

Sample size and composition

The study involved approximately 30,000 volunteers (one third women, two thirds men), ranging between 15 to 65 years. The people were from diverse working environments, ranging from employees of 52 large companies and the public service in Munster and the Northern Ruhr Area of Germany.

Duration

The study extended for a duration of 6 to 8 years. The rate of participation was an average of 60%.

Collection of data

Data was collected from questionnaires, examination and interview by physician. Measurement of blood pressure, resting electrocardiogram, case history and family history, collection of blood sample after 12-hour fast, etc. were also used to obtain data and to account for these factors on the results.

Valid and reliable results

Since the data was obtained by following standard procedures and the sample size was large, the data should be reliable. The data were analysed statistically and the results were compared with other similar studies to check the validity of the results.

Edexcel IAL from 2018 Specification 1.17: understand why people's perception of risks are often different from the actual risks, including underestimating and overestimating the risks due to diet and other lifestyle factors in the development of heart disease

Calculation of risk

Risk describes the probability that an event will happen. Probability means the chance or likelihood of the event, calculated mathematically.

In 2005, 335969 people died due to Coronary Heart Diseases (CHD). The total UK population at that time was 60 209 408, so we can calculate the average risk in a year of someone in the UK dying from Coronary Heart Diseases (CHD) as:

$$\text{Risk} = (\text{Number of events} / \text{total entities}) \times 100$$

$$\text{Risk} = (335969 / 60\ 209\ 408) \times 100 = 0.55\%$$

Risk perception

Risk perception is **one's opinion of the likelihood of risk.** *Risk perception* (how we judge risk) is an important concept in healthy living. Human perceptions of risk are not very accurate. Our judgments about risks are based upon several factors.

The table below shows the perception of risk by the public, collected from a survey on risk assessment in the United States. It also shows the actual risk for each factor, which is calculated by experts. The possible reason for overestimation or underestimation is also stated in the last column.

Scale: 1 represents the lowest risk and 20 represent the highest risk.

Factor	Experts	Public	Perception by public and reason for estimation
Nuclear Power	1	20	Overestimated – as it is **dreaded** and **not natural**. It is also **Involuntary (Not under their control) and unfair.** The **actual risk** (chance or probability) is a lot lower.
Motor Vehicles	2	1	Underestimated – as it is **voluntary** to drive a motor vehicle and the person feels in **control** of the activity.
Handguns	3	4	Overestimated – as it is **dreaded** by many.
Smoking	4	2	Underestimated – as it is **voluntary** and the **effects are not immediate**.
Fire fighting	5	18	Overestimated – as it is **dreaded** and **not natural**. The **actual risk** (chance or probability) is a lot lower.
Alcoholic Beverages	6	3	Underestimated – as it is **voluntary** and the **effects are not immediate**.
Private Aviation	7	12	Overestimated – as it is **dreaded** and **Involuntary (Not under their control)**. The **actual risk** (chance or probability) is a lot lower.
Police Work	8	17	Overestimated – as it is **dreaded** and **Involuntary (Not under their control)**. The **actual risk** (chance or probability) is a lot lower.
Pesticides	9	8	Underestimated – as it is **voluntary** and the **effects are not immediate**.
Surgery	10	5	Underestimated – as it is **voluntary.**

Risk perception and CVDs

It is a well-known fact that a diet high in saturated animal fats, sodium chloride and consumption of large amounts of alcohol are factors which increase the risk of developing cardiovascular diseases. Lack of physical activity and smoking are other lifestyle factors that could also increase the risk of suffering from cardiovascular diseases. However, we tend to ignore the risk associated with these components of our diet and lifestyle because of the underestimation or overestimation of the risk.

Self-assessment (Chapter 7)

Step 1: Visit www.teachingtube.com
Step 2: Sign up for the free course "Edexcel AS and AL Bio assessments"
Step 3: Attempt the chapter Quiz and get instant results.
Step 4: Download the assessment worksheet and marking scheme too.

CHAPTER EIGHT

CHOLESTEROL AND CARDIO-VASCULAR DISEASES

Edexcel IAL from 2018 Specification 1.18: (i) be able to analyse data on the possible significance for health of blood cholesterol levels and levels of high-density lipoproteins (HDLs) and low-density lipoproteins (LDLs)
(ii) know the evidence for a causal relationship between blood cholesterol levels (total cholesterol and LDL cholesterol) and cardiovascular disease (CVD)

Data analysis – example 1

The data on age-specific death rates from coronary heart disease between 1980 and 2005 is shown in the table below.

Death rates per 100 000 of population of the United Kingdom due to coronary heart disease

Age	35 – 44		45 – 54		55 – 64		65 – 74	
	Men	Women	Men	Women	Men	Women	Men	Women
1980	56	9	270	50	733	215	1621	688
1990	37	6	159	33	536	179	1352	594
2000	19	5	92	20	291	84	823	347
2005	19	4	73	16	204	54	558	225

The data can be interpreted to draw the following conclusions.

Conclusion one: The average risk of death due to coronary heart disease increases as the population grows older. This is substantiated by the fact that the death rate increases with age for both men and women, during the entire period of the study. For example, in 1980 the death rate for 35 to 44-year-old men is 56 per 100 000 of the population. This number rises in 65 to 74-year-old men to 1621 per 100 000 of the population. The same trend is seen for women. This may be due to the fact that arteries get less elastic as the person ages. This increases the blood pressure and increases the risk of heart diseases.

Conclusion two: The number of deaths from coronary heart disease between 1980 and 2005 decreases. For example, the death rate in 45 to 54-year-old women decreases from 50 per 100 000 of the population in 1980 to 16 per 100 000 of the population in 2005. A similar trend is seen for other age groups as well. This may be due to improvements in their diet, which may contain less saturated fat due to increased awareness of risks and lifestyle factors. The awareness factor may also contribute to increased physical activity or exercise to minimise the risk of coronary heart diseases. Improved screening and diagnostic tools help to identify people who are at greater risk and allows them to make lifestyle and diet changes. Drugs like statins, are used as a preventive measure to keep the LDL levels low and reduces the risk of Coronary Heart Diseases (CHD).

Data analysis and conflicting evidence – example 2

The graphs below show the incidence of angina per 100 000 adults determined in two UK studies, the 4th National Study of Morbidity Statistics from General Practice (1991 /1992) and One general practice in Oxford (1989 / 1991).

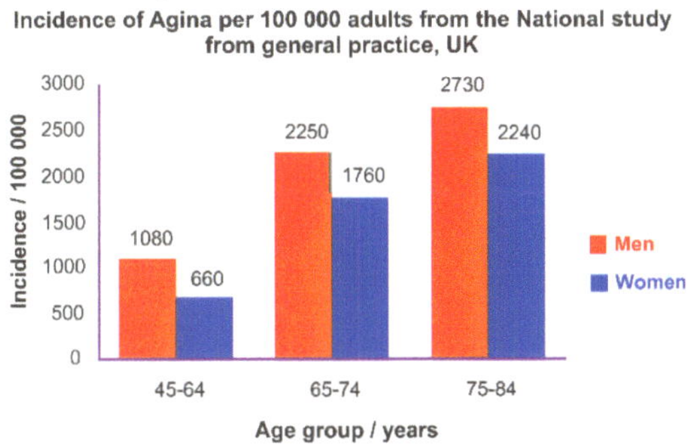

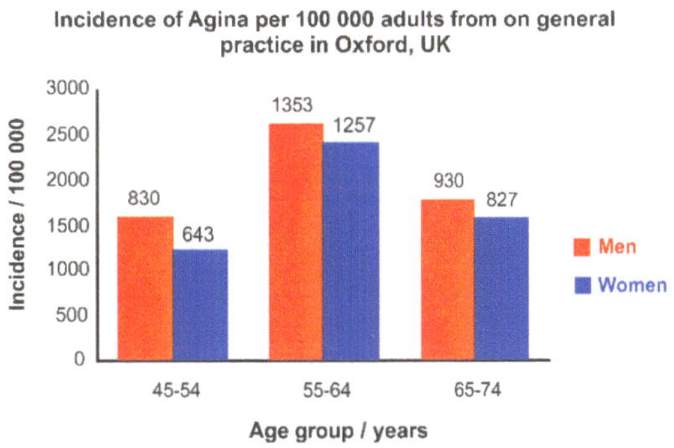

Figure 8.1: Data from the 4th National Study of Morbidity Statistics from General Practice (1991 /1992) and One general practice in Oxford (1989 / 1991)

The following conclusions can be drawn from the studies.

Conclusion one

At all ages, men have a higher incidence of death from coronary heart disease than do women. The difference between men and women decreases in the older age groups, as shown in fig.10.3 and 10.4.

This may be because men have a greater genetic predisposition. Men may be consuming more alcohol and smoking more than women, which increase the formation of atheroma.

Oestrogen in women has a protective effect on the heart as it reduces the LDL levels in blood. So, women are at lower risk before menopause. After menopause, the protective effect of oestrogen is not present, so the difference in risk is lower in older age groups.

Conflicting evidence

National Study from General Practice (1991 – 1992)		Study of one general practice in Oxford (1989 – 1991)
This study shows that the risk of angina for both men and women increases with age from 45 to 84	**but**	the Oxford study shows an increase followed by a decrease after the age of 64 for both men and women
Also, males in the National GP study aged **45-64 had a 1 in 93 chances** of having angina, those between **65 and 74 had a 1 in 45 chances** of angina. **How risk is calculated?** Between the age of 45 – 64, 1080 men out of 100,000 suffer angina symptoms, so it means that 1 out of every 92.6 or 93 men suffered the symptoms. $100{,}000 \div 1080 = 92.6$ Or $(1080 \div 100{,}000) \times 100 = 1.08\%$ of the population suffered the symptoms.	**but**	in the Oxford study men aged **45–64 had a 1 in 46 chances** of having angina, those between **65 and 74 had a 1 in 107 chances** of angina. This pattern is repeated for women too. **Risk calculation.** Between the age of 65 – 74, 930 men out of 100,000 suffer angina symptoms, so it means that 1 out of every 107 men suffered the symptoms. $100{,}000 \div 930 = 107$ Or $(930 \div 100{,}000) \times 100 = 0.93\%$ of the population suffered the symptoms.
The differences in the two studies may be due to the different year of the study, the size of the sample studied (one is a national study, the other draws data from a single general practice) or the methods used in the study.		

Edexcel IAL from 2018 Specification 1.19: **understand how people use scientific knowledge about the effect of diet, including obesity indicators, such as body mass index and waist-to-hip ratio, exercise and smoking to reduce their risk of coronary heart disease**

Obesity is a risk factor for coronary heart diseases (CHD). Thus, one must consume a diet which will prevent the person from getting obese. The energy content of the diet must be balanced out with the physical activity level and Basal Metabolic Rate of the individual. Obesity and risk of CHD can be measured by obesity indicators. The figure 8.2 shows the effects of Obesity.

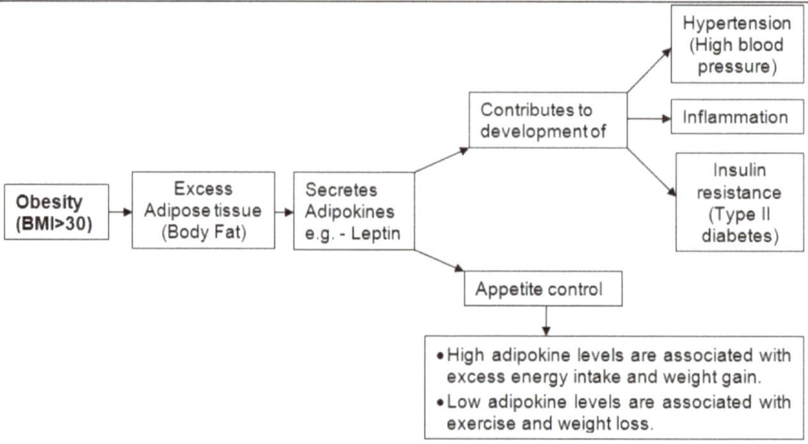

Figure 8.2: Obesity

Obesity indicators – and objective means of determining obesity

Overweight and obesity increase the risk of CHD. As well as being an independent risk factor, obesity is also a major risk factor for high blood pressure, raised blood cholesterol, diabetes and impaired glucose tolerance'.

The adverse effects of excess weight are more pronounced when fat is concentrated in the abdomen. This is known as **central or abdominal obesity** and is assessed using the waist to hip ratio. The World Health Organization's World Health Report 2002 estimated that over 7% of all disease burden in developed countries was caused by raised body mass index (BMI), and that around a third of CHD and ischaemic stroke and almost 60% of hypertensive disease in developed countries was due to overweight.

More recently the INTERHEART case-control study estimated that 63% of heart attacks in Western Europe and 28% of heart attacks in Central and Eastern Europe were due to abdominal obesity (a high waist to hip ratio), and those with abdominal obesity were at over twice the risk of a heart attack compared to those without. This study also found that abdominal obesity was a much more significant risk factor for heart attack than BMI.

1. Waist-to-Hip Ratio

The waist to hip ratio is one of the most used clinical applications of girth measurements. This assessment is important *because* there is a correlation between chronic diseases and fat stored in the midsection. The waist-to-hip ratio can be computed by dividing the waist measurement by the hip measurement by doing the following

1. Measure the *smallest* part of the waist, without drawing the stomach inward, as in figure 8.3.

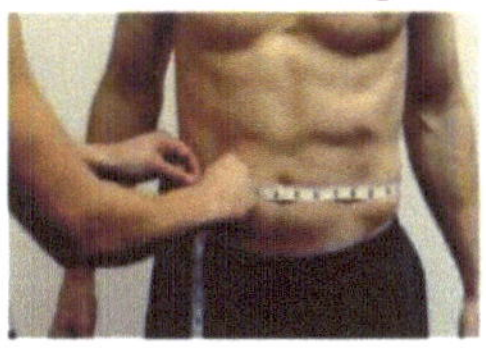

Figure 8.3: Measuring the waist

2. Measure the largest part of the hip, as in figure 8.4.

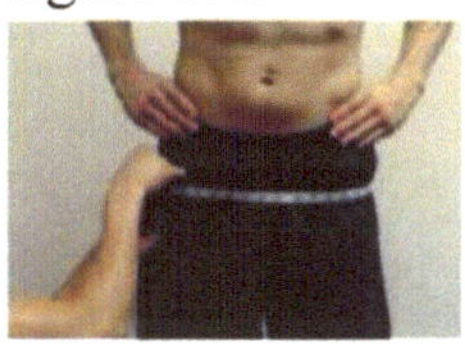

Figure 8.4: Measuring the hip

3. Compute the hip-to-waist ratio by dividing the waist measurement by the hip measurement.
4. For example. if a client's waist measures 30 inches and his or her hips measure 40 inches, divide 30 by 40 for a hip-to-waist ratio of 0.75. A ratio above 0.80 for women and above 0.95 for men may put these individuals at risk for several diseases.

2. Body Mass Index (BMI)

The body mass index (BMI) is a quick and easy method for determining whether a person's weight is appropriate for his or her height. To assess weight relative to height, divide body weight (in kilograms) by height (in meters squared) or, Kg/m^2.

$$BMI = Weight\ (in\ kg) / Height^2\ (in\ m)$$

It has been shown that obesity-related health problems increase when a person's BMI exceeds 25. The BMI cut offs listed below apply to 'normal' healthy adults. *It is not appropriate for pregnant women, or for use in some medical conditions or with children. They may also be inappropriate for athletes and some ethnic groups (as BMI does not distinguish between fat and fat-free mass). For example, a BMI over 27.5 in an Asian person has been estimated by the World Health Organisation to carry the same health risk as a BMI of 30 in a white Caucasian person.* It is more difficult to assess obesity in children, so special charts have been developed which consider growth, gender and age.

This table shows the way different BMIs are classified and the relationship between BMI and risk of associated diseases

Body Mass Index (kg/m^2)	Classification	Risk of disease associated with excess weight
Less than 20	Underweight	Low (but increased risk of other clinical problems)
Over 20 to 25	Desirable or healthy range	Average
Over 25 to 30	Overweight	Increased
Over 30 to 35	Obese (Class I)	Moderate
Over 35 to 40	Obese (Class II)	Severe
Over 40	Morbidly or severely obese (Class III)	Very severe

Keep in mind the BMI and hip-to-waist measurement does not consider percent body fat or lean body mass. Thus, one may possess a high BMI or hip-to-waist ratio score but have a low percentage of body fat and high percentage of lean body mass, indicating their risk for obesity-related health problems may not be as high as their results may show.

Exercise

The easiest way to increase physical activity is by incorporating more physical activity into daily routines, e.g. walking or cycling instead of driving (particularly for short journeys) and taking up more active hobbies such as gardening. The benefits of exercise are shown in figure 8.5.

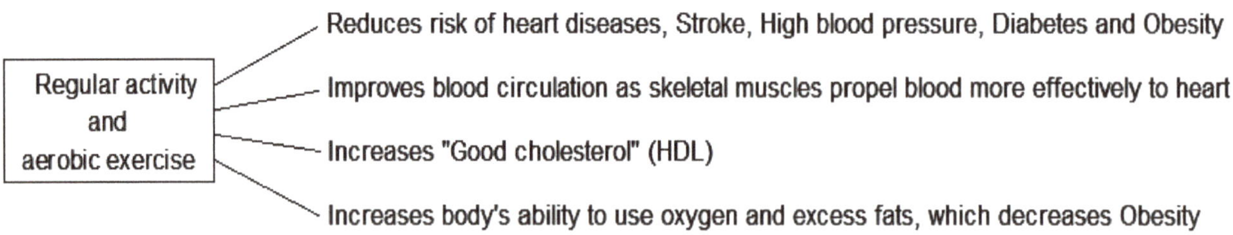

Figure 8.5: Benefits of exercise

Caution: If you have a high risk of heart disease, a history of heart disease, diabetes, or other serious health problem, get your doctor's guidance before beginning or increasing an exercise program.

Smoking

The effects of smoking have already been explained. The pie chart in figure 8.6 shows the estimated average annual number of smoking-attributable deaths in the United States during 2000 through 2004 by specific causes. The data shows that 32.03% of the deaths are caused by stroke and ischaemic heart disease, associated with cigarette smoking.

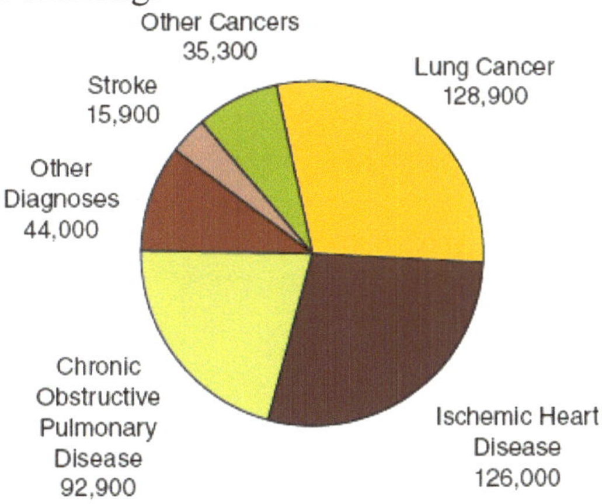

Figure 8.6: Smoking and CVDs

Diet and CVDs: Energy basics

- Energy is needed by the body to stay alive, grow, keep warm and be active.
- Energy is provided by food and drink. It comes from the fat, carbohydrate, protein and alcohol.
- Food and drinks provide different amounts of energy.
- **Different people need different amounts of energy**.
- Some activities use more energy than others.
- **To maintain body weight, it is necessary to expend as much energy as is derived from food; to lose weight, energy must exceed intake**.
- Energy can be measured in either ***joules*** or ***calories***. One calorie is equivalent to 4.2 joules.

How much energy do individuals need?

The actual amount of energy needed will vary from person to person and depends on their *basal metabolic rate* (BMR) and their level of physical activity.

Basal metabolic rate

The basal metabolic rate can be thought of as the rate at which a person uses energy to maintain the basic functions of the body, *i.e.* an adult will use around 4.6 kJ (1.1 kcal) every minute. BMR is measured when a person is at complete rest and varies from person to person both within a population group and between different groups. Infants and young children have a proportionately high BMR for their size due to their rapid growth and development. Men usually have a higher BMR than women since they tend to have more muscle. Older adults usually have a lower BMR than the young since the amount of muscle tends to decrease with age. The BMR accounts on average for about three quarters of an individual's energy needs.

Energy balance

When the diet provides **more energy** than is needed, the excess is stored as fat and the person **'puts on weight'**. The excess energy is stored in the body as fat in *adipose* tissue, as depicted in figure 8.8. When the diet provides **less energy** than is needed, the person **'puts loses weight'**, as depicted in figure 8.7. When the diet provides **the same amount of energy** that is being used, the person's weight **remains constant**, as depicted in figure 8.9.

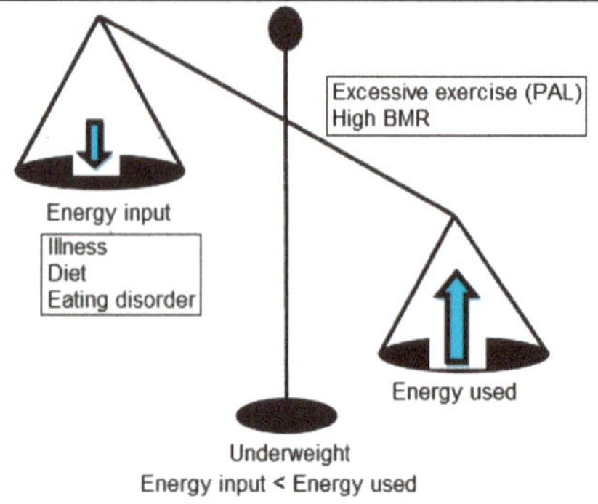

Figure 8.7: Energy imbalance

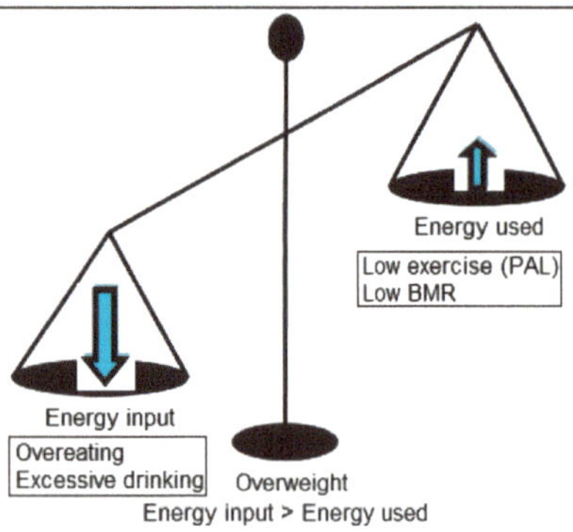

Figure 8.8: Energy imbalance

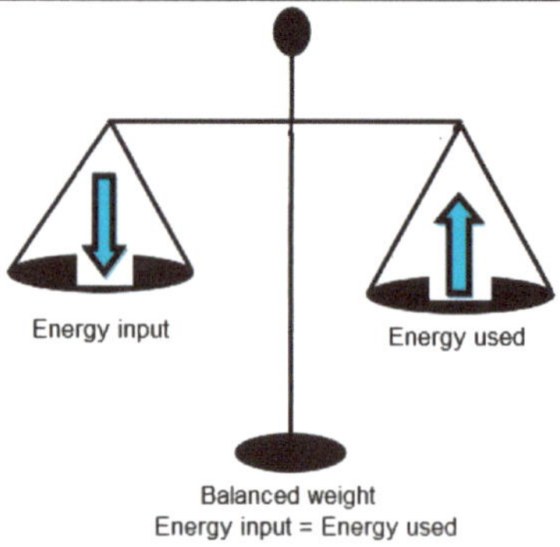

Figure 8.9: Energy imbalance

The table below shows the energy in a meal and the physical activities that could use up the energy consumed **(Energy in versus energy out).**

Energy consumed = 400 KJ	**Physical activity to burn 400 kJ of energy**	
White rice boiled (68g)	**Boys**	**Girls**
Milk (205ml)	Running (16.5 minutes)	Running (17.5 minutes)
Margarine (13g)	Walking (30.5 minutes)	Walking (32.5 minutes)
Milk Chocolate (18g)	Cycling (16.5 minutes)	Cycling (17.5 minutes)
Wholemeal bread (44g)	Swimming (23.5 minutes)	Swimming (25.0 minutes)
Tomatoes (548g)		
Oranges (peeled) (253g)		

Dietary Reference Values (DRVs)

Dietary Reference Values (DRVs) comprise a series of estimates of the **amount of energy (EAR)** and **nutrients (RNI)** needed by different groups of healthy people in the UK population.

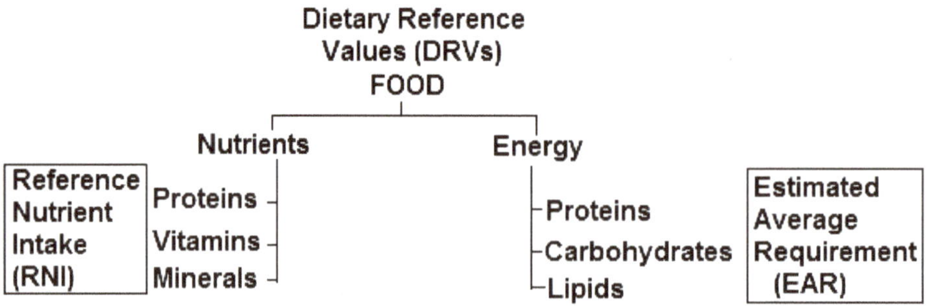

Figure 8.10: DRVs for various types of food

Energy: The **energy recommendations** for diet are provided by Estimated Average Requirement Values **(EAR).** It is estimated by considering the physical activity levels (PAL) and the (Basal Metabolic Rate) BMR of the population. The data is then analysed and a reference value is estimated.

Energy EAR = BMR x Physical Activity Level (PAL)

The table below shows the estimated average requirement (EAR) of energy for various groups in the UK population.

EAR - MJ/day (kcal/day)					**EAR - MJ/day (kcal/day)**				
Age	*Males*		*Females*		*Age*	*Males*		*Females*	
	(MJ)	*(kcal)*	*(MJ)*	*(kcal)*		*(MJ)*	*(kcal)*	*(MJ)*	*(kcal)*
0-3 mo	2.28	(545)	2.16	(515)	11-14 yr	9.27	(2220)	7.72	(1845)
4-6 mo	2.89	(690)	2.69	(645)	15-18 yr	11.51	(2755)	8.83	(2110)
7-9 mo	3.44	(825)	3.20	(765)	19-50 yr	10.60	(2550)	8.10	(1940)
10-12 mo	3.85	(920)	3.61	(865)	51-59 yr	10.60	(2550)	8.00	(1900)
1-3 yr	5.15	(1230)	4.86	(1165)	60-64 yr	9.93	(2380)	7.99	(1900)
4-6 yr	7.16	(1715)	6.46	(1545)	65-74 yr	9.71	(2330)	7.96	(1900)
7-10 yr	8.24	(1970)	7.28	(1740)	74+ yr	8.77	(2100)	7.61	(1810)

Nutrients: the nutrient requirements are provided in the form of Reference Nutrient Intake values (RNIs). Nutrients are obtained from **proteins, vitamins and minerals.**
The table below shows the Reference Nutrient Intake (RNI) for Vitamin C for different groups. **Nutrients: the nutrient requirements are provided in the form of Reference Nutrient Intake values (RNIs).**

The table below shows the Reference Nutrient Intake (RNI) for Vitamin C for different groups.

Nutrient	Age group	RNI (mg / day)
	0 – 6 months	40
Vitamin C	7 – 12 months	50
	1 – 3 years	15
Also known as Ascorbic acid	4 – 8 years	25
(Dehydro ascorbic acid – DHA)	Males	
	9 – 13 years	45
	14 – 18 years	75
(Function: Cofactor for reactions requiring reduced copper or iron metalloenzyme and as a protective antioxidant)	19 – 30 years	90
	31 – 50 years	90
	50 – 70 years	90
	> 70 years	90
	Females	
	9 – 13 years	45
	14 – 18 years	65
	19 – 30 years	75
	31 – 50 years	75
	50 – 70 years	75
	> 70 years	75

Note: Similar tables of RNI exist for other vitamins, minerals and proteins.

The advantage of publishing these values is that it helps to plan diets for individuals or groups of people in the population. It can also be useful for assessing the diets of people. The labels on food products have the calorific values to help people to calculate how much of energy they need to consume to keep healthy.

For example, if you are a 17-year-old boy, living in the UK, you would need to consume

- **2755 kcal of energy (EAR) and**
- **75 mg of vitamin C (RNI) every day.**

You could then look at food labels to check for the calorific value of the food you consume. It should ideally be about 2755 kcal per day. If you consume less than the recommended value, then you could be underweight and may suffer from deficiency disorders. If you consume more than the recommended value, you may become obese. Likewise, food labels also give the amount of vitamin C per unit volume of juice. You could calculate how much of a fruit juice you would need to consume to get the 75 mg of vitamin C per day (RNI).

The EAR for women who become pregnant increases by 0.8 MJ/day (200 kcal/day) but only in the final three months. Although energy is needed for the growth of the fetus and to enable fat to be deposited in the mother's body, pregnant women can compensate for these extra demands by becoming less active and using energy more efficiently.

Breastfeeding mothers have increased requirements for energy, but this will depend on the amount of milk produced, the fat stores that have accumulated during pregnancy and the duration of breastfeeding.

Self-assessment (Chapter 8)

Step 1: Visit www.teachingtube.com
Step 2: Sign up for the free course "Edexcel AS and AL Bio assessments"
Step 3: Attempt the chapter Quiz and get instant results.
Step 4: Download the assessment worksheet and marking scheme too.

CHAPTER NINE

TREATMENT OF CARDIO-VASCULAR DISEASES

Edexcel IAL from 2018 Specification 1.20: know the benefits and risks of treatments for cardiovascular disease (CVD) (antihypertensives, statins, anticoagulants and platelet inhibitors)

Treatment of atherosclerosis

Treatments for atherosclerosis may include lifestyle changes, medicines, and medical procedures or surgery.

Drugs for treatment of CVD

Many medicines are used to treat patients with cardiac disease. However, individual patients may respond differently to these medications. The effectiveness of a medication may also change with time. For this reason, doctors often prescribe different medications to control, hypertension, for example, until they find a medicine or combination of medicines that works for a given patient. Commonly used medications include: antihypertensives, statins, anticoagulants and platelet inhibitory drugs.

1. Antihypertensives

Antihypertensives are used to treat high blood pressure (hypertension). The most common types include ARB, ACE inhibitors, diuretics, beta blockers and calcium channel blockers.

Figure 9.1 shows the mechanism by which changes in blood pressure is regulated.

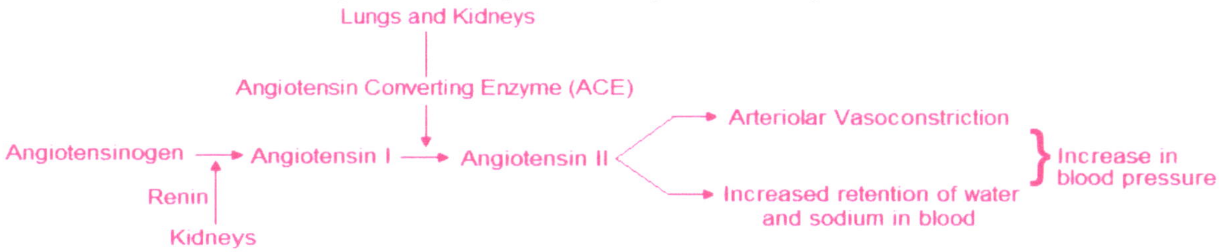

Figure 9.1: Regulation of blood pressure

The table below summarises the effects of antihypertensives and the risks and benefits.

Mode of action	Benefits	Risks
a. Angiotensin Receptor Blockers (A2RBs) Angiotensin is a substance produced by the liver and kidneys collectively. It causes vessels to constrict by binding with receptors on smooth muscles of the arteries. This elevates the blood pressure. Angiotensin receptor blockers (ARBs) block the action of angiotensin. So, ARBs lower the blood pressure by preventing constriction of the blood vessels.	ARBs are used for controlling high blood pressure, treating heart failure, and preventing kidney failure in people with diabetes or high blood pressure.	Dizziness, headache, drowsiness, diarrhoea, abnormal taste sensation (metallic or salty taste), and rash.
b. ACE Inhibitors ACE (angiotensin converting enzyme - produced by the lungs and kidneys) is used in the formation of angiotensin. **ACE (angiotensin converting enzyme) inhibitors** are used to treat high blood pressure and weakened heart muscles. ACE inhibitors slow (inhibit) the activity of the enzyme (angiotensin converting enzyme), which produces angiotensin. As a result, the blood vessels dilate, and blood pressure is reduced. In addition, these drugs decrease sodium and water retention.	ACE inhibitors are used for controlling high blood pressure, treating heart failure, and preventing kidney failure in people with diabetes or high blood pressure.	Skin rash, hay-fever-like symptoms (sneezing, blocked or runny nose, itchy eyes), swelling of your sinuses (sinusitis), sore throat or dry cough, feeling sick or vomiting, indigestion

c. Diuretics Diuretics act on the kidneys to help rid the body of excess water and sodium and consequently lower the blood pressure as well.	Diuretics are commonly used to treat high blood pressure (hypertension), congestive heart failure (CHF), and water retention and swelling.	Weakness, Muscle cramps, Skin rash, Diarrhoea, muscle cramps and dizziness.
d. Beta-Blocking Drugs Beta-adrenergic receptors are receptors found on many cells in the body. These receptors bind with the hormone adrenaline. This allows adrenaline to cause constriction of the arteries and to increase the heartbeat rate. Beta-blockers block the beta-adrenergic receptors (beta-adrenergic blocking drugs) and slow the heart rate and decrease blood pressure. They do this by blocking the effects of adrenaline on the body's beta receptors. This slows the nerve impulses that travel through the heart, thus easing the heart's pumping action and widening blood vessels.	Beta-blockers are used to treat a variety of conditions including high blood pressure (hypertension), chest pain (angina), arrhythmias, congestive heart failure (CHF), and mitral valve prolapse. These drugs are especially useful to treat angina caused by stress or psychological shock.	Fatigue, depression, memory loss, dizziness, diarrhoea or constipation.
e. Calcium Channel Blocking Agents Also called calcium antagonists or calcium blockers, these medications affect the movement of calcium into the cells of the heart and blood vessels. As a result, they relax blood vessels and increase the supply of blood and oxygen to the heart, while reducing its workload. Calcium-channel blockers can cause a big drop in blood pressure. This can cause dizziness while standing up. So, after taking a dose, patients need to get up slowly and stay next to your chair or bed until the dizziness wears off.	They are prescribed to treat chest pain (angina), high blood pressure (hypertension) and some irregular heartbeats (arrhythmias). They are not usually prescribed for people with heart failure or other structural damage to the heart.	Calcium-channel blockers can also slow the heart down and cause headaches, constipation, flushing and fluid retention.

2. Cholesterol lowering drugs – plant statins

Cholesterol is a lipid. It is an important component of cell membranes and it is used in the synthesis of several sex hormones. About half of our daily cholesterol requirement is absorbed from the gut. The remainder is synthesised in the liver. The figure 9.2 shows the formation of cholesterol in the liver.

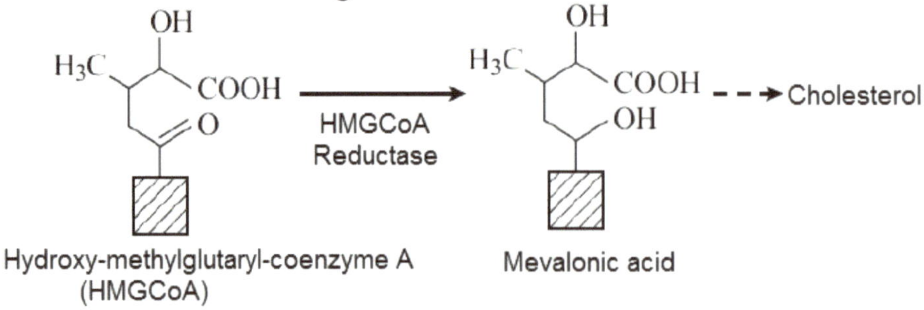

Figure 9.2: Formation of cholesterol in the liver

Statins block the enzyme in the liver that is responsible for making cholesterol. This enzyme is called hydroxy-methylglutaryl-coenzyme A reductase (HMG-CoA reductase for short). Scientifically, **statins are referred to as HMG-CoA reductase inhibitors**.

Benefits of statins

Statins are used to lower the blood cholesterol (LDL) levels and reduce the risk of CVD. Doctors often prescribe statins if diet, weight loss, and exercise don't work to sufficiently lower a person's cholesterol (LDL).

Risks of using statins

- Muscle and joint aches (most common)
- Nausea
- Diarrhoea
- Constipation
- **Liver damage (rarely)**

3. Anticoagulants

Anticoagulants ("blood thinners") are medicines that delay the clotting of blood. They inhibit clot formation by blocking the action of clotting factors which are part of the blood clotting cascade.

Examples are heparin and warfarin.

- Warfarin in **inhibits the manufacture of prothrombin** in the body. Low prothrombin levels make the blood clot less easily.
- Heparin **inhibits the formation of thrombin**.

4. Platelet inhibitors

Antiplatelet medicines keep blood clots from forming by preventing blood platelets from sticking together. Antiplatelet drugs interact with platelets, which are fragments of cells that trigger blood clotting. These drugs block platelets from aggregating into harmful clots. Example, aspirin and clopidogrel.

Benefits of anticoagulants and platelet inhibitory drugs

- Although blood clotting is essential to prevent serious bleeding in the case of skin cuts. However, clots inside the blood vessels block the flow of blood to major organs and cause heart attacks and strokes. Anticoagulants and platelet inhibitors reduce the ability of the blood to form clots and lower the risk of CVDs.
- Anticoagulants and platelet inhibitors are also used to reduce the chance of blood clots forming during open heart surgery or bypass surgery. Low doses of these drugs may be given to prevent blood clots in patients who must stay in bed for a long time after certain types of surgery.

Risks of using anticoagulants and platelet inhibitory drugs

- Because anticoagulants affect the blood's ability to clot, they can increase the risk of severe bleeding and heavy blood loss. It is thus essential to take these drugs exactly as directed and to see a physician regularly as long as they are prescribed.
- The risks of taking aspirin are well known - it irritates the stomach lining and causes bleeding in the stomach which can become serious.

CHAPTER TEN

GAS EXCHANGE SURFACES

Edexcel IAL from 2018 Specification 2.1: (i) know the properties of gas exchange surfaces in living organisms (large surface area to volume ratio, thickness of surface and difference in concentration)
(ii) understand how the rate of diffusion is dependent on these properties and can be calculated using Fick's Law of Diffusion
(iii) understand how the structure of the mammalian lung is adapted for rapid gaseous exchange

All living organisms need to respire. Respiration requires oxygen and releases carbon dioxide. If the oxygen supply is less than the demand, anaerobic respiration begins to form lactic acid. Lactic acid accumulates in the cells and decreases the pH of the cytoplasm. This low pH will denature the enzymes, carrier and channel proteins in the cell and the cell will malfunction. So, oxygen must be continuously supplied to the cells and carbon dioxide must be removed. Cells also require ATP for active transport, protein synthesis, DNA replication and other metabolic activity. ATP is produced by respiration. So, a continuous supply of oxygen must be provided for the cells to remain metabolically active.

Gas exchange in single celled organisms

In single celled organisms like amoeba, oxygen and carbon dioxide are exchanged directly between the environment and the surface of the organism, as shown in figure 10.1. This method of exchange, by diffusion alone, is sufficient to sustain the metabolic activity of amoeba because of the following reasons

1. Amoeba has a **large surface area to volume ratio** because of its small size
2. A very **small diffusion distance**. This means that all the surfaces are very near to the center. So, gases can rapidly diffuse from all around the cell to the center.
3. Respiration constantly uses up oxygen and produces carbon dioxide. This ensures that a **concentration gradient is maintained continuously**, and oxygen can diffuse into the cell while carbon dioxide diffuses out.

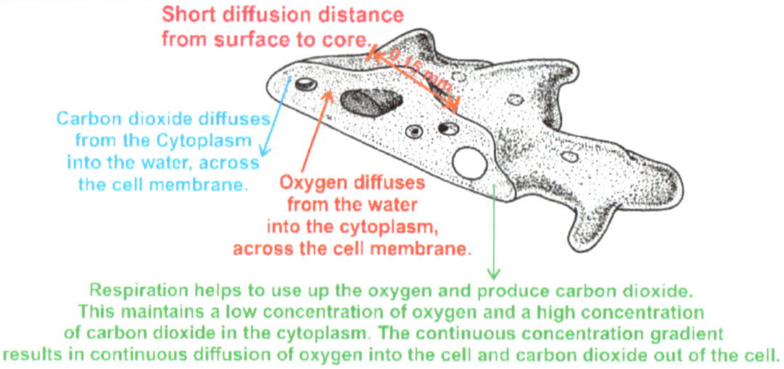

Figure 10.1: Gas exchange in amoeba

Gas exchange in larger, multi-cellular organisms

Larger organisms like plants and animals have a low surface area to volume ratio and large diffusion distance. Oxygen and carbon dioxide cannot be **rapidly transported** across the body surface. These organisms have developed special respiratory surfaces, like leaves, lungs and gills. The respiratory surfaces have three features in common.

1. **large surface area**
2. **short diffusion distance**
3. **A mechanism to maximize the concentration gradient.**

The relationship between these factors and the rate of diffusion is given by Fick's law, as shown below

$$\text{Rate of diffusion} \propto \frac{\text{Surface area x concentration difference}}{\text{Distance}}$$

Fick's law has been used to derive a few formulae which help to calculate the rate of diffusion.

Equation 1

$J = -D\frac{dC}{dx}$	Where, J is the flux → Flow / area / time with the unit as (atoms / m^2 / s) D is the diffusion coefficient with the unit cm^2/s dC is the difference in concentration of particles unit is particles / mm^4 dx is the distance moved by the particles during diffusion

Application of Fick's law to calculate diffusion

An *E. coli* cell averages about 2 μm in length. If D for the amino acid alanine in cells is 0.00021 cm^2/s and difference in concentration is 0.2 particles / mm^4. Find the flux.

$$J = -D\frac{dC}{dx}$$

$= -0.00021 \times (0.2 / 0.000001)$

$= 42$ particles of alanine /m^2/s

If the value of D and concentration gradient is the same, then calculate the flux for *S. cerevisiae* (baker's yeast) which measures about 9 μm.

$$J = -D\frac{dC}{dx}$$

$= -0.00021 \times (0.2 / 0.0000045)$

$= 9.3$ particles of alanine / m^2 / s

The flux or rate of diffusion is 4.5 times more in E.coli than in Yeast cells. This is because of the larger size of the yeast cells. So, the larger the cell the slower is the flux or rate of diffusion in the cell.

Remember for cells the diffusion distance is the distance from the cell membrane to the center of the cell, as shown in figure 10.2.

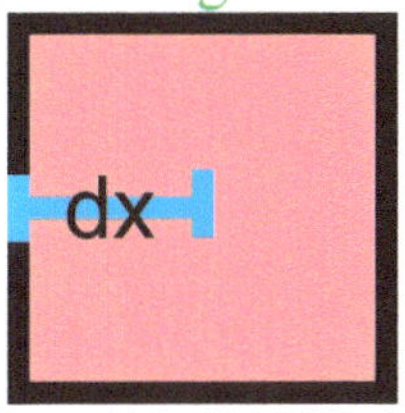

Figure 10.2: (dx) Diffusion distance in cells

Equation 2

$T = \frac{1}{2D} x^2$	Where, T is the time required to diffuse D is the diffusion coefficient with the unit cm^2/s x is the distance moved by the particles during diffusion

Application of Fick's law to calculate diffusion

If D = 0.00001 cm^2/s (for oxygen through water), how long would it take the oxygen to diffuse 0.01 cm below the surface of a still pond?

$$T = \frac{1}{2D} x^2$$

$= (1/2 \times 0.00001) \times 0.01^2$

= 5 seconds

And if the oxygen were to diffuse from the surface of the pond to a depth of 1cm it would take 50,000 seconds.

Any fish at a depth of 1 cm below the water surface would die from oxygen debt if they had to rely on diffusion to bring in oxygen. Luckily there are water currents to distribute gases rapidly.

Gas exchange in plants

Plants absorb Carbon dioxide for photosynthesis and oxygen for respiration. The feature that help rapid diffusion of these gases in plants are listed below.

1. **large surface area:** Numerous leaves and large number of air spaces in the mesophyll layer.
2. **short diffusion distance:** Thin flattened lamina, as shown in figure 10.3, decreases the diffusion distance.
3. **A mechanism to maximize the concentration gradient:** respiration and photosynthesis help to maintain a concentration gradient.

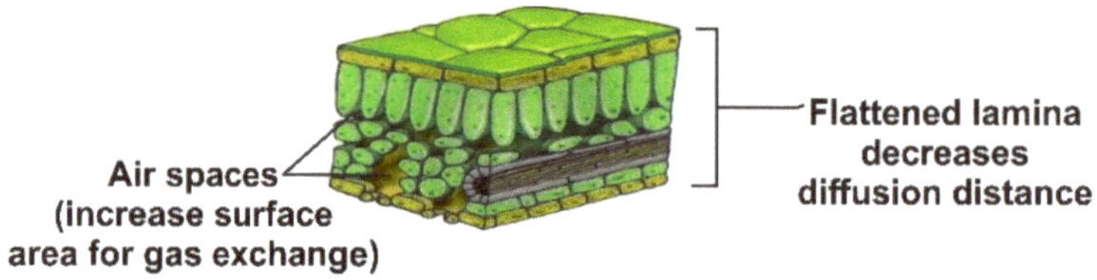

Figure 10.3: Diffusion of gases in a leaf

The human lungs

The human lungs are in the thoracic cavity. The lungs are surrounded by the ribs, diaphragm, vertebral column, sternum and intercostal muscles as shown in figure 10.4. Each lung is supplied with air from the bronchi, as shown in figure 10.5.

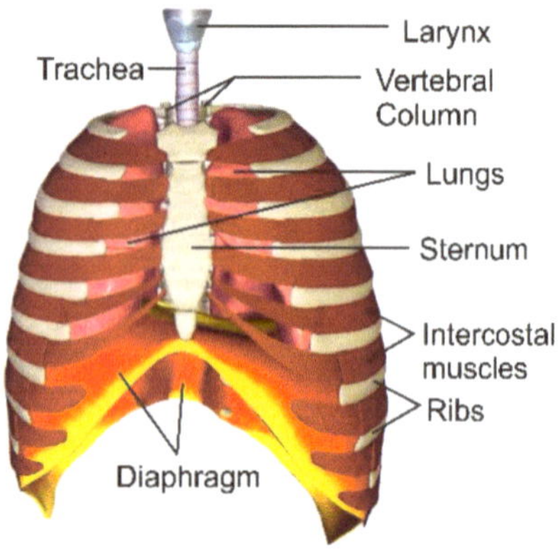

Figure 10.4: The thoracic cavity

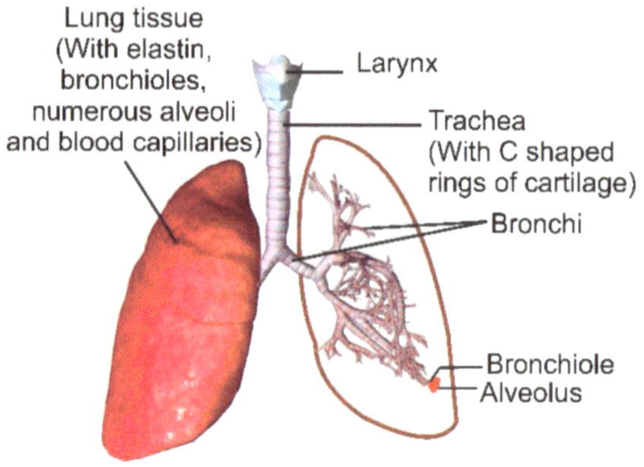

Figure 10.5: Structure of the lungs

The human lungs are composed of millions of tiny sacs, alveoli, dispersed into a matrix of connective tissue called elastin. The elastin matrix also contains numerous blood capillaries, as depicted in figure 10.6.

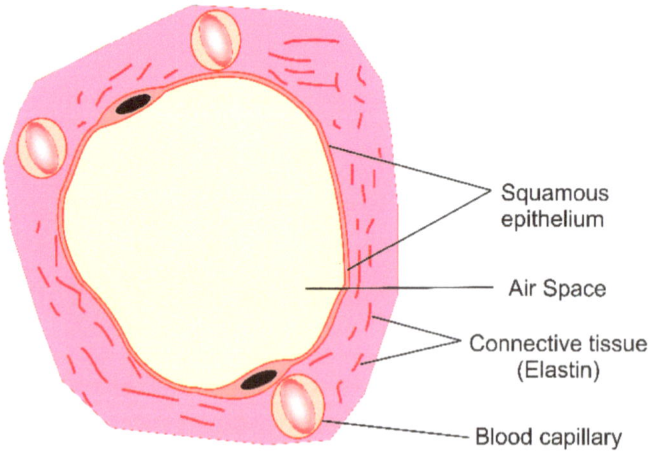

Figure 10.6: An alveolus surrounded by blood capillaries

The surfactant is a monolayer of phospholipid molecules present on the inner surface of the alveolus as shown in figure 10.7. The lung **surfactant prevents the alveoli from collapsing due to cohesive forces of water molecules during exhalation.** This makes inflation of the alveoli easy. Some babies born without surfactant in the lungs die from exhaustion. This is because they need to spend too much energy to inflate the lungs. This is called the respiratory distress syndrome. The surfactant also **reduces the surface tension of water on the inner lining of the alveoli,** making it easier for gases to pass across the walls.

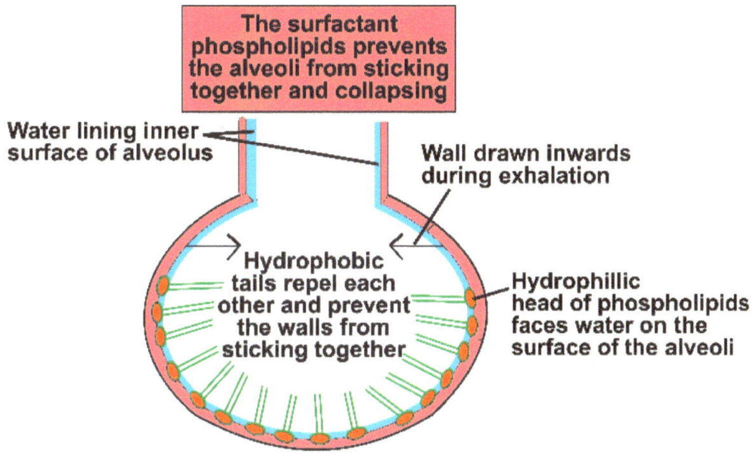

Figure 10.7: Surfactant in the lungs

In humans the gas exchange organ system is the respiratory or breathing system.
The main features are:

- The walls of the alveoli are composed of a single layer of flattened epithelial cells, as are the walls of the capillaries, so gases need to diffuse through just two thin cells. This decreases the **diffusion distance**.

- The **steep concentration gradient** across the respiratory surface is maintained in two ways: by blood flow on one side and by air flow on the other side.

- There are millions of alveoli and blood capillaries, which provide **an extremely large surface area** for gases to be exchanged.

- The flow of air in and out of the alveoli is called **ventilation** and has two stages: **inspiration** (or inhalation) and **expiration** (or exhalation). Lungs are not muscular and cannot ventilate themselves, but instead the whole **thorax** moves and changes size, due to the action of two sets of muscles: the **intercostal muscles** and the **diaphragm**, as shown in figure 10.8 and 10.9.

During inspiration / inhalation	During exhalation / expiration
• The external intercostal muscles contract. • The internal intercostal muscles relax. • The ribs and sternum are pulled outwards and upwards. • The diaphragm contracts and becomes flattened. • Volume of thoracic cavity increases causing pressure to become less than atmospheric pressure, so that air rushes into lungs.	• The internal intercostal muscles contract. • The external intercostal muscles relax. • The ribs and sternum pulled inwards. • Diaphragm relaxes and becomes dome shaped. • Volume of thoracic cavity decreases. • Pressure in lungs becomes greater than atmospheric pressure, so air is pushed out of the lungs.

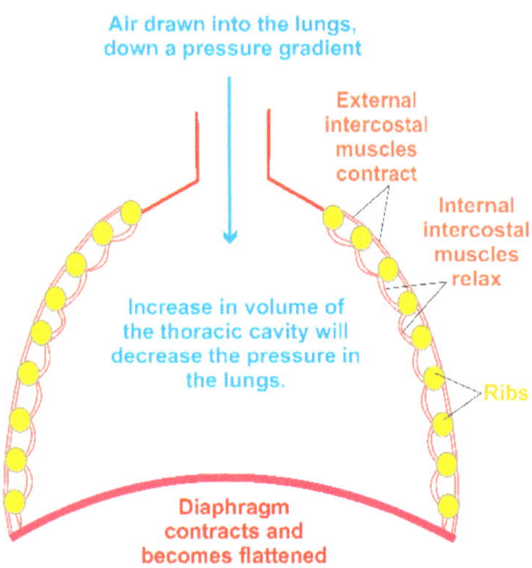

Figure 10.8: Changes during inhalation

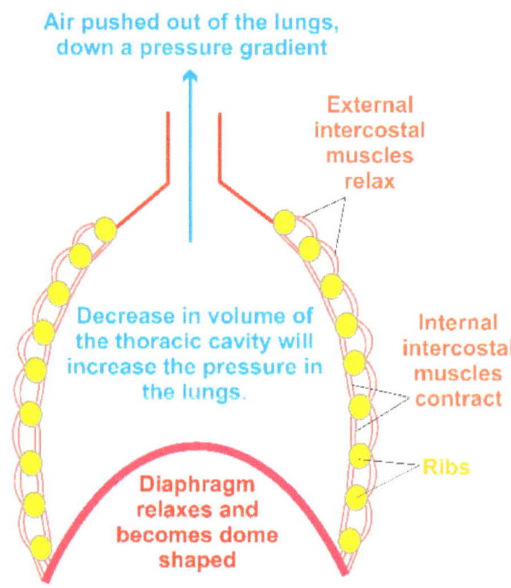

Figure 10.9: Changes during exhalation

- The C-shaped rings of cartilage help to keep the trachea and bronchi open always. If the cartilage is removed, then the trachea and bronchi will collapse, and breathing would not be possible.
- Goblet cells in the trachea secrete sticky mucus to trap pathogens and dust particles from the air. The mucus and pathogens are then swept out of the trachea by microscopic cilia, as shown in figure 10.10. The contaminated mucus is swallowed, and the pathogens are destroyed by hydrochloric acid in the stomach.

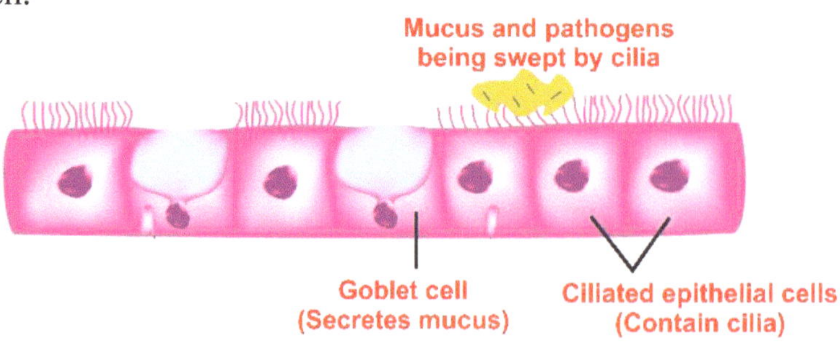

Figure 10.10: Cilia lining the trachea

Self-assessment (Chapter 10)

Step 1: Visit www.teachingtube.com
Step 2: Sign up for the free course "Edexcel AS and AL Bio assessments"
Step 3: Attempt the chapter Quiz and get instant results.
Step 4: Download the assessment worksheet and marking scheme too.

CHAPTER ELEVEN

STRUCTURE OF THE CELL MEMBRANE

Edexcel IAL from 2018 Specification 2.2: (i) know the structure and properties of cell membranes
(ii) understand how models such as the fluid mosaic model of membrane structure are interpretations of data used to develop scientific explanations of the structure and properties of cell membranes

Structure and properties of the cell membrane

The cell membrane is a selectively permeable membrane which helps to prevent the passage of some substances and allows the passage of other substances. This helps to maintain the appropriate composition of the cytoplasm.

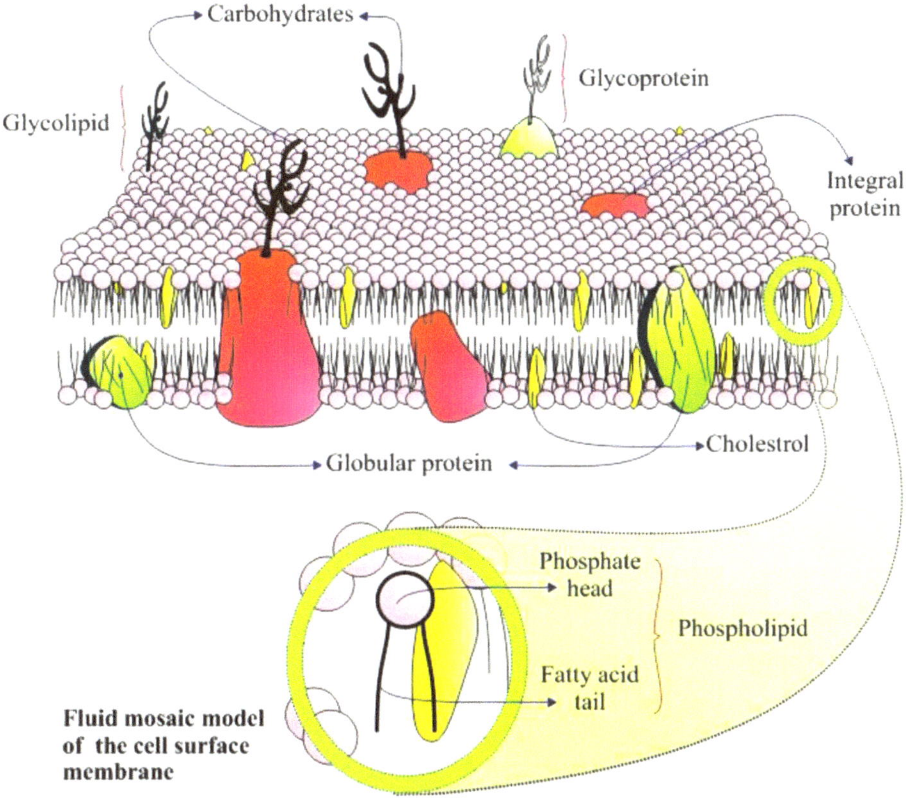

Figure 11.1: Cell membrane structure

Phospholipid bilayer

The phospholipid bilayer is **permeable to non-polar molecules** like CO_2, O_2 and **lipid soluble molecules** like steroids. The hydrophobic **fatty acid tails form a barrier to polar molecules and ions** like glucose, amino acids, Na^+ ions, K^+ ions, Ca^{2+} ions, Cl^- ions, etc.

Proteins

Proteins embedded in the phospholipid bilayer allow **specific** polar molecules and ions to pass across the membrane. Hence, they are often referred to as transporter proteins - **channel proteins and carrier proteins**. Some of the proteins (extrinsic proteins) act as enzymes, recognition sites and electron carriers. Other proteins may serve as **receptors** for hormones and neurotransmitters, play a role in **cell adhesion** by sticking to membrane proteins on other cells and act as **antigens** for cell recognition.

Glycoproteins and glycolipids

Branched chains of carbohydrates maybe attached to some phospholipid molecules, forming **glycolipids** or to proteins, forming **glycoproteins**. This is collectively called the **glycocalyx**. The glycocalyx is always **found only on the outer surface** of the cell membrane. The glycocalyx provides **chemical protection** to the cell membrane and plays an important role in **cell recognition and adhesion**.

Cholesterol

Cholesterol makes the membranes less fluid and more stable. Unsaturated fatty acid tails in the phospholipid bilayer will increase the fluidity of the membrane. Scientists began building models of the cell membranes decades before membranes were first seen with the electron microscope. An early study of red blood cell membranes revealed that the membrane was composed mainly of **lipids** and **proteins**. The red blood cells were a good choice for the research as they do not have internal membranes or organelles.

Gorter and Grendel model

In 1924, Gorter and Grendel proposed that the cell membrane must actually be **a phospholipid bilayer**. They calculated the **area of the red blood cell membrane** and then extracted the lipids and dissolved it in petroleum ether. The lipids were then allowed to spread into a layer one molecule thick on a surface of water and **the area was measured**.

Total surface area of RBC membrane / mm^2	Surface area of lipid monolayer on water surface/mm^2	Ratio
4.7×10^{-7}	**9.4×10^{-7}**	**1:2**

Since, phospholipid molecules have non-polar hydrophobic fatty acid tails and a polar, phosphate hydrophilic head they arrange themselves into a bilayer. The hydrophobic tails faced away from water on the membrane surfaces while the hydrophilic heads were attracted to water on the membrane surfaces, as shown in fig. 11.2.

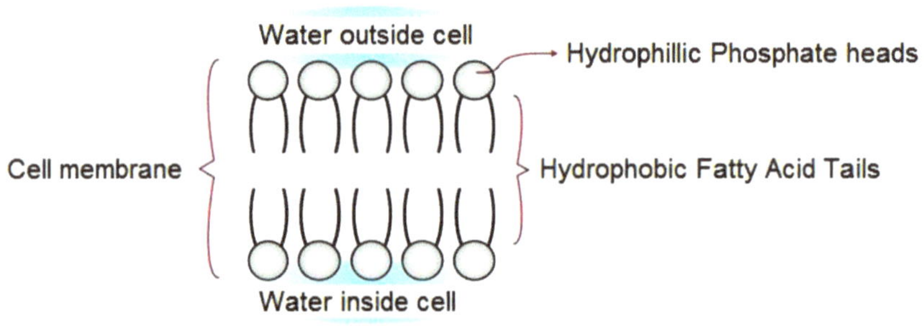

Figure 11.2: Formation of a phospholipid bilayer

With the conclusion that a phospholipid bilayer was the main fabric of a membrane, the next question was where to place the proteins. It was also clear at that time that a simple phospholipid bilayer was too weak and unstable to hold the cell contents.

Davson and Danielli - sandwich model

In 1935, Davson and Danielli, first proposed the **sandwich model**. It suggested that the stability of the membrane could be accounted for if the **phospholipid bilayer was sandwiched between two layers of proteins.** The first electron micrographs in 1950s supported the Davson-Danielli model, as shown in figure 11.3.

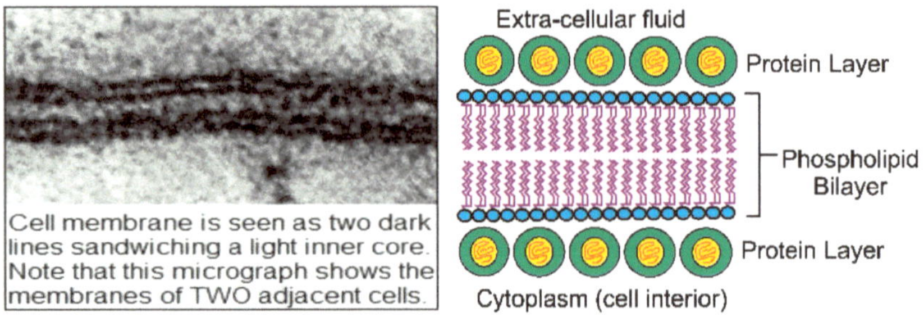

Figure 11.3: Davson-Danielli Sandwich model of the cell membrane

However, many cell biologists in the 1970s recognized **two main problems** with the Davson-Danielli sandwich model.

i. This model assumes that all membranes are identical - this was known to be false:

For example, the cell membrane is 7 to 8 nm in thickness and has a three-layered structure in electron micrographs, whereas the inner mitochondrial membrane has a thickness of only 6nm and appeared to have bead like structures. Mitochondrial membranes also have a substantially greater percentage of proteins than the plasma membranes.

ii. The membrane proteins would be exposed to hydrophilic environments on all sides (from the phospholipids and from the water of the cytoplasm). This is not a stable configuration: Analysis of proteins in the membrane revealed that the membrane proteins were amphipathic – having both hydrophilic and hydrophobic regions. This makes the membrane proteins relatively insoluble in water. If such proteins were layered on the surface of the membrane, their hydrophobic regions would be in an aqueous environment.

Singer and Nicolson - Fluid-Mosaic model

Finally, in 1972, S.J. Singer and G. Nicolson proposed the **Fluid-Mosaic model** of the cell membrane. They proposed that the **Cell surface membrane** is made up of a phospholipid bilayer, with **proteins randomly embedded into the phospholipid bilayer**, as shown in figure 11.4.

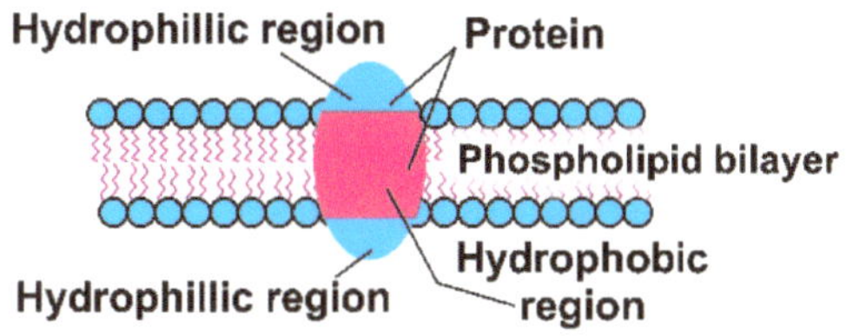

Figure 11.4: Singer-Nicolson Fluid mosaic model of the cell membrane

The fluid mosaic model was able to explain the beaded nature of mitochondrial membranes and could also explain the variable composition of proteins in the membranes.

The evidence for this was confirmed by the freeze fracture technique. A cell is frozen in liquid nitrogen and fractured with a knife. The fracture often splits the membrane along its hydrophobic interior. The two layers of phospholipids are split, and the proteins are exposed as beads or bumps, demonstrating that the proteins are embedded in the bilayer, as shown in figure 11.5.

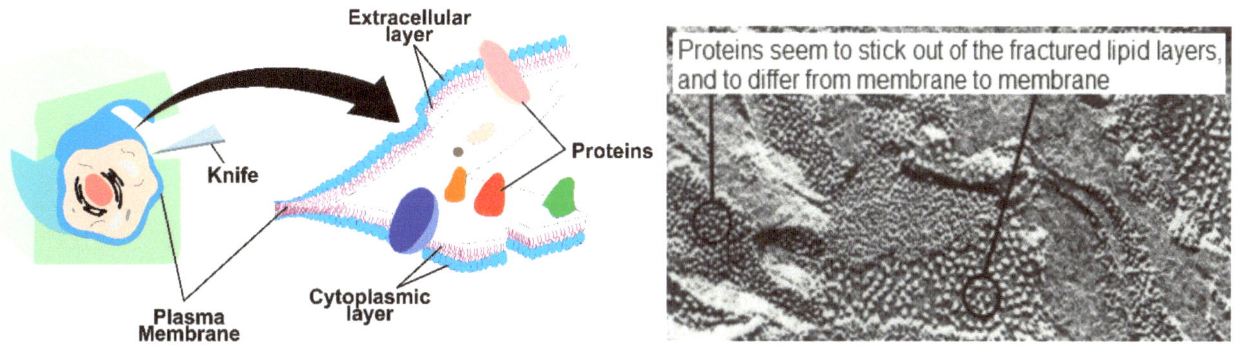

Figure 11.5: Freeze etching provides evidence for the fluid mosaic model

Meaning of Fluid-Mosaic

Fluid means that **molecules can change places** within the membrane. **Mosaic** means that **proteins are randomly embedded** in to the phospholipid bilayer.

The following experiment shows evidence of the fluid mosaic model.

The plasma membrane proteins of a mouse cell and a human cell were labelled with two different dyes, as shown in figure 11.6. The two cells were then fused and observed under a microscope. The proteins were found to have changed places and rearranged themselves randomly. This proves that the membrane is a **fluid mosaic**. The phospholipid molecules can also change places laterally within the membrane at the rate of 10 million times a second.

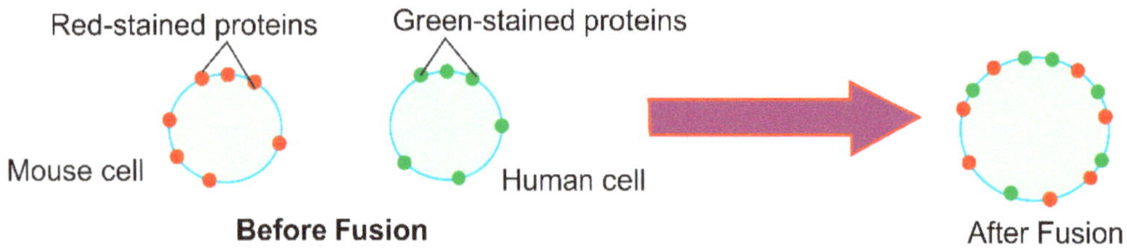

Figure 11.7: Evidence of fluidity in the cell membrane

Discovery of aquaporins

It was previously thought that water can diffuse through the phospholipid bilayer because of its small size and low polarity. However, many scientists were not convinced as this could not explain the differential permeability of different cell membranes, specially the cells in the kidney. In 2003, Peter Agre discovered special protein channels called **aquaporins** (Water channels) in the cell membranes. It is now known that water passes through the cell membrane through the phospholipid bilayer and the aquaporins as well. The aquaporins allow the cell to control water movement by opening or closing of aquaporins.

> Edexcel IAL from 2018 Specification 2.3: CORE PRACTICAL 3
> **Investigate membrane properties including the effect of alcohol and temperature on membrane permeability.**

Experimental procedure

Step one: Cut beetroot discs of uniform surface area to volume ratio.

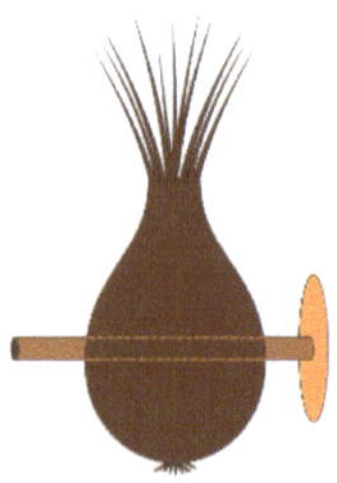

Figure 11.8 – Use a cork-borer to obtain cylinder of tissue from beetroot

Figure 11.9 - Expel cylinder of beetroot from the cork borer

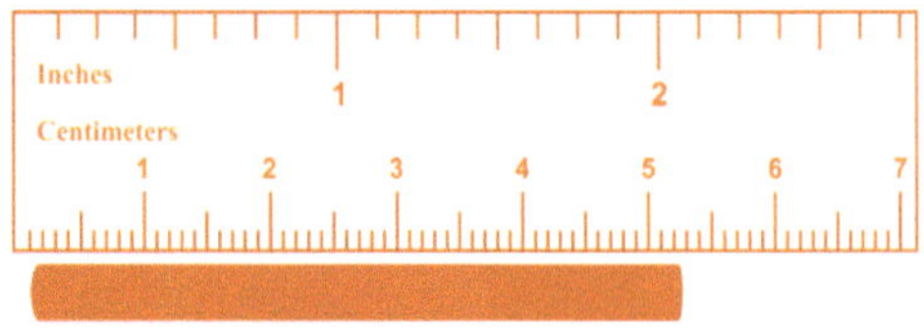

Figure 11.10 - Place the beetroot cylinder on a tile and cut into 3 mm wide discs

Place all the discs in a small beaker and wash under a running tap for at least 15 minutes. This removes cell debris and pigment from the surface of the discs.

Step two: Equilibration of temperature

Gently pick 6 beetroot discs with forceps, one by one. Place the discs in a test tube and place it in the water bath for three minutes.

Transfer 6 cm^3 of distilled water into another test tube by using a graduated pipette. Place this test tube into the same water bath as the discs for a few minutes.

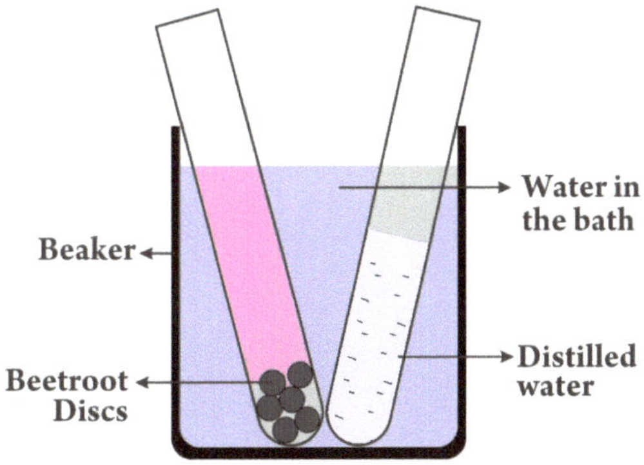

Figure 11.11 – Equilibration of Temperature

Then pour the distilled water into the test tube containing the discs.
This is referred to as equilibration of temperature because the beetroot discs and distilled water reach the same temperature as the water bath before it is mixed.

Step Three: Treatment of beetroot discs

Leave the discs in the test tubes for 20 minutes.

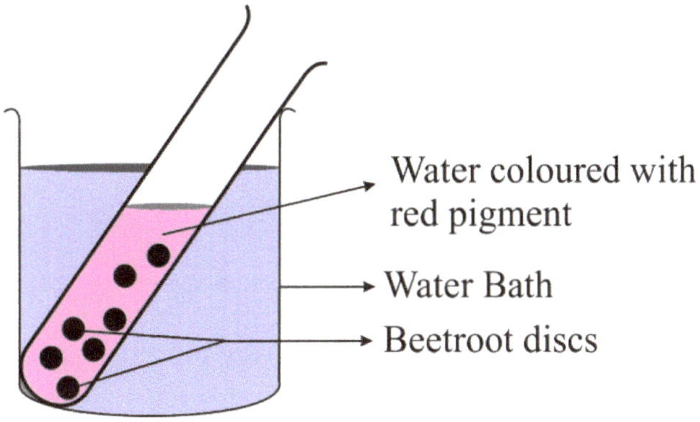

Figure 11.12 – Temperature treatment

Step four: Estimate the intensity of the red pigment by using a colorimeter.

Shake the tubes to distribute the colour uniformly and transfer 5 cm^3 of the coloured solution from the test tube into a cuvette.

Set the colorimeter to absorbance mode (the other mode is transmission – do not use this mode as it will give an inverse relation to intensity of colour).

Set the colorimeter filter to a blue filter for this experiment.

Fill a cuvette with distilled water and set the set the absorbance to zero. This is called as the blank reading and is essential for calibration of the colorimeter. Then remove the cuvette with distilled water and place the cuvette with the red pigment into the colorimeter and record the absorbance.

The absorbance is proportional to intensity of the red colour, which is an indication of the extent of membrane damage or permeability.

Working of the colorimeter – Cuvette contains pigment solution

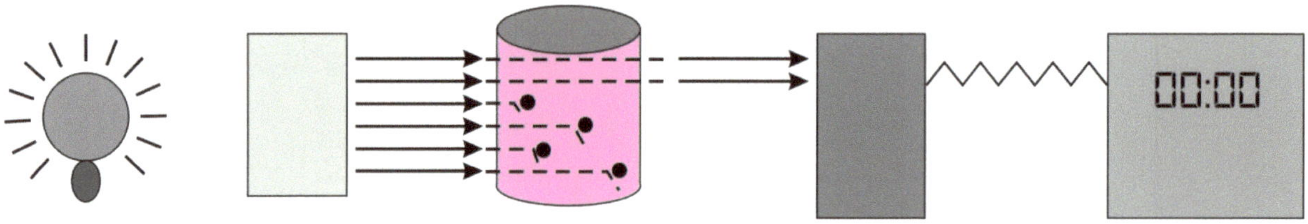

Figure 11.13 – Working of a Colorimeter

Step five: Repeat the experiment at different temperatures while keeping all other factors unchanged.

Core practical for effect of ethanol on permeability of cell membrane

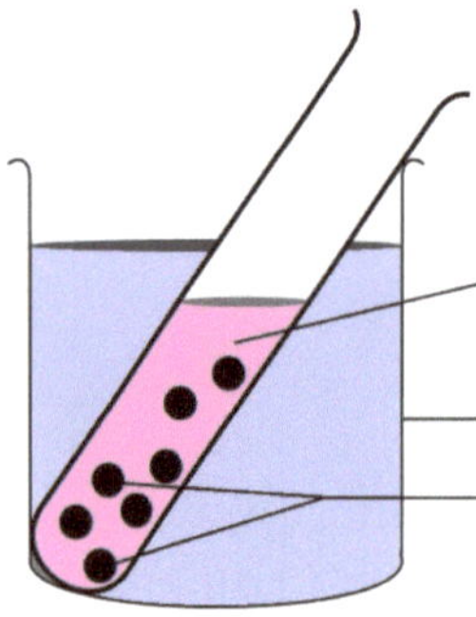

Figure 11.14 – Effect of ethanol on cell membrane

The experiment can be repeated in the same way as describe for effect of temperature on cell membrane. However, the test tube should contain ethanol instead of distilled water and the temperature should be kept constant. **Ethanol dissolves the phospholipid bilayer.**

Self-assessment (Chapter 11)
Step 1: Visit www.teachingtube.com
Step 2: Sign up for the free course "Edexcel AS and AL Bio assessments"
Step 3: Attempt the chapter Quiz and get instant results.
Step 4: Download the assessment worksheet and marking scheme too.

CHAPTER TWELVE
TRANSPORT ACROSS CELL MEMBRANES

Edexcel IAL from 2018 Specification 2.4: understand what is meant by osmosis in terms of the movement of free water molecules through a partially permeable membrane, down a water potential gradient

Osmosis – transport of water across the cell membrane

Osmosis is the net movement of water molecules from a region of higher water potential to a region of lower water potential across a partially permeable membrane, as shown in figure 12.1.

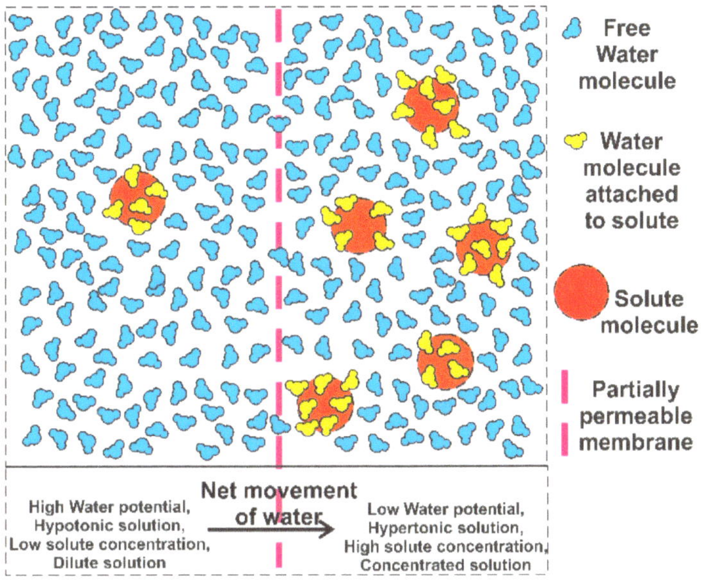

Figure 12.1: Osmosis

Water Potential

Water Potential is the tendency of water molecules to move away from a solution. Osmosis can be quantified using water potential, so we can calculate which way water will move, and how fast. Water potential (ψ , the Greek letter psi, pronounced "sy") is a measure of the water potential for movement in a solution. It is measured in units of pressure (Pa, or usually kPa), and the rule is that water always moves by osmosis from less negative to more negative water potential (in other words it's a bit like gravity potential or electrical potential). 100% pure water has $\psi = 0$, which is the highest possible water potential, so all solutions have a water potential less than zero ($\psi < 0$, i.e. a negative number).

There are two factors which affect the water potential

a) Solute concentration: Pure water has a water potential of 0 KPa. Adding solute into pure water will decrease its water potential. For example, a solution containing 17g of sucrose in $1dm^3$ of water has a water potential of -130KPa. A solution containing 35g of sucrose in $1dm^3$ of water would have a water potential of -260 KPa. This simply means that the water molecules will have a lesser tendency to move away from a more concentrated solution (with a low ψ).

b) Pressure on both side of the membrane: Consider a plant cell placed in pure water. Water enters the cell by osmosis down a water potential gradient. However, this does not go on forever. The inward movement of water will stop when the cell becomes turgid, even though a concentration gradient still exists. This is because the cell wall exerts a pressure on the water molecules and decreases its tendency (water potential) to enter the cells.

The formula for calculation of water potential is: $\psi = \psi_s + \psi_p$

Where, ψ is the water potential.

ψ_s is the solute potential

ψ_p is the pressure potential.

Figure 12.2 shows the movement of water by osmosis between three plant cells.

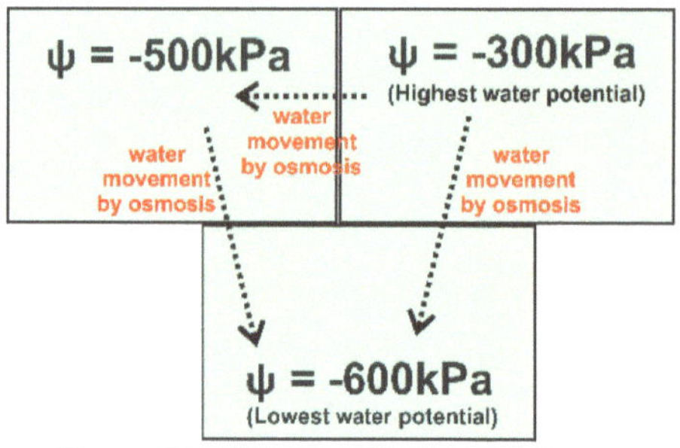

Figure 12.2: Direction of movement of water

Cells and Osmosis

Osmolarity is a measure of *solute concentration*, as defined by the number of osmoles of a solute per litre of solution (osmol/L).

Solutions may be loosely categorised as hypertonic, hypotonic or isotonic according to their relative osmolarity.

- Solutions with a relatively *lower* osmolarity are categorised as **hypotonic** (low solute concentration ⇒ loses water)
- Solutions that have the *same* osmolarity are categorised as **isotonic** (same solute concentration ⇒ no net water flow)
- Solutions with a relatively *higher* osmolarity are categorised as **hypertonic** (high solute concentration ⇒ gains water)

The effects of these solutions on cells are shown in figure 12.3.

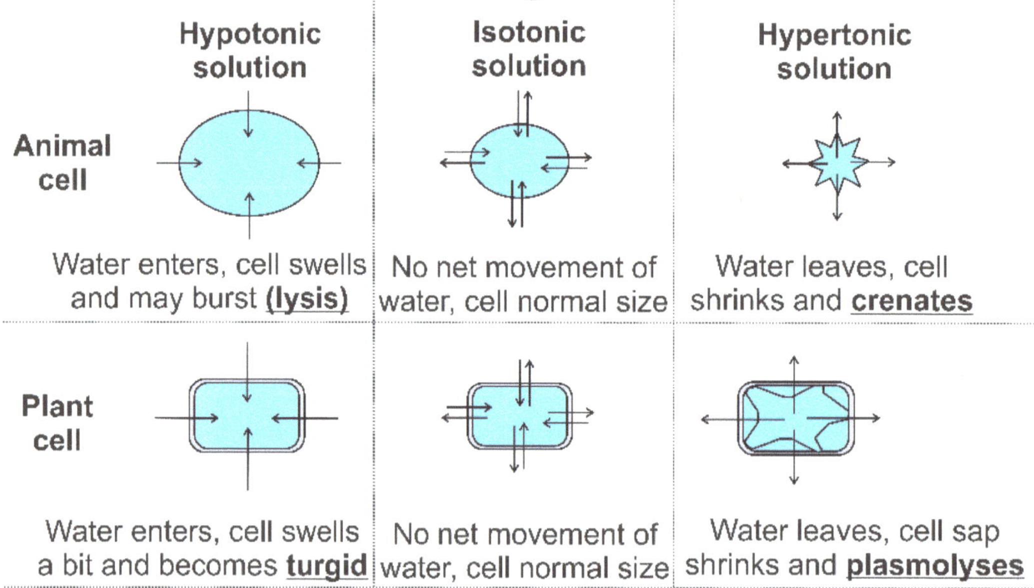

Figure 12.3: Osmosis in cells

When 2 fresh raw eggs with their shells removed with acid are placed into a strong sucrose solution (hypertonic) and distilled water (hypotonic). Water enters the egg in water (endosmosis) causing it to swell and water leaves the egg in sucrose causing it to shrink (exosmosis).

These are problems that living cells face all the time. For example:

- Simple animal cells (protozoans) in fresh water habitats are surrounded by a hypotonic solution and constantly need to expel water using contractile vacuoles to prevent swelling and lysis.
- Cells in marine environments are surrounded by a hypertonic solution and must actively pump ions into their cells to reduce their water potential and so reduce water loss by osmosis.
- Young non-woody plants rely on cell turgor for their support, and without enough water they wilt. Plants take up water through their root hair cells by osmosis and must actively pump ions into their cells to keep them hypertonic compared to the soil. This is particularly difficult for plants rooted in salt water.

Edexcel IAL from 2018 Specification: RECOMMENDED ADDITIONAL PRACTICAL Investigate tissue water potentials using plant tissue and graded concentrations of a solute.

Procedure

1. Use 1M sucrose solution and distilled water to prepare a series of 10 cm3 sucrose solutions in boiling tubes of different concentrations: 1M, 0.8M, 0.6M, 0.4M 0.2M and 0.0M. Label the boiling tubes.
2. Use a cork borer to obtain cylinders of potato tissue with the same diameter. Cut them all to the same length of 5 cm. It is important to work quickly to avoid loss of water through evaporation as this would lower the water potential of the tissue.
3. Immerse two cylinders of potato tissue in each tube and cover the tubes with sealing film.
4. Leave the set up for one hour.
5. Remove the cylinders from each tube. Measure the length of the cylinders, and calculate the percentage change in length using the formula below:

$$\text{\% change in length} = \frac{\text{final length} - \text{initial length}}{\text{initial length}} \times 100\%$$

6. Find the mean percentage change in length of the cylinders at each concentration.
7. Plot a graph of the mean percentage change in length (vertical axis) against the concentration of sucrose solution (horizontal axis).

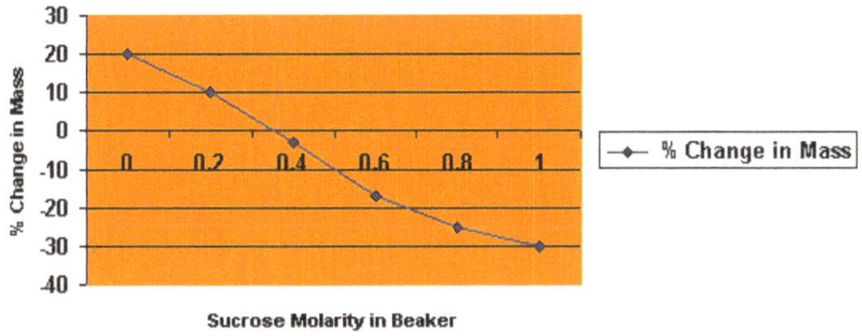

Figure 12.4: Graph showing the effect of osmosis on plant tissue

8. From the graph, determine the concentration of sucrose solution which causes no change in length of the tissue.
9. The water potential of the potato tissue can be expressed in terms of the molarity of sucrose solution that causes no change in length of the tissue.

Edexcel IAL from 2018 Specification 2.5: (i) understand what is meant by passive transport (diffusion, facilitated diffusion), active transport (including the role of ATP as an immediate source of energy), endocytosis and exocytosis
(ii) understand the involvement of carrier and channel proteins in membrane transport

Transport of other substances across the cell membrane

The cell membrane is a selectively permeable membrane which helps to prevent passage of some substances and allows the passage of other substances. This helps to regulate the appropriate composition of the cytoplasm. The Two-dimensional structure of the cell surface membrane and its permeability is illustrated in fig. 12.5.

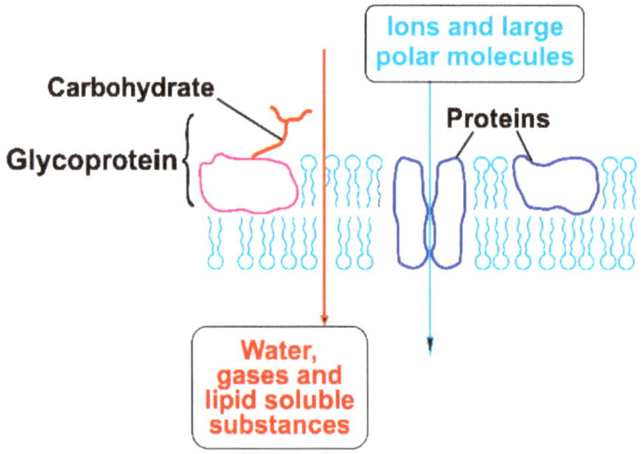

Figure 12.5: Selective permeability of the cell membrane

Molecules move in and out of cell across the selectively permeable cell membrane by four basic processes, namely, diffusion, facilitated diffusion, active transport and bulk transport (exocytosis and endocytosis).

Simple Diffusion

Diffusion is the **net** movement of particles from a region of its higher concentration to a region of its lower concentration, due to the kinetic energy of the molecules. The particles ***move in both directions*** across the membrane, but the rate of movement of particles from higher to lower concentration is greater than the movement in the opposite direction. Each type of molecule or ion moves down its own concentration gradient independent of other molecules. For Example: O_2 and CO_2 diffuse in different directions at the same time, in the alveoli.

Substances that can be exchanged across the phospholipid bilayer by diffusion

- O_2 and CO_2 are nonpolar, small molecules which can diffuse rapidly across the phospholipid bilayer.
- The nonpolar fatty acid tails of the phospholipid bilayer serve as a barrier to ions and large polar molecules, like glucose, amino acids, Na^+, Cl^-, etc.
- Steroid hormones and glycerol are lipid soluble and can easily diffuse across the phospholipid bilayer of the membrane.

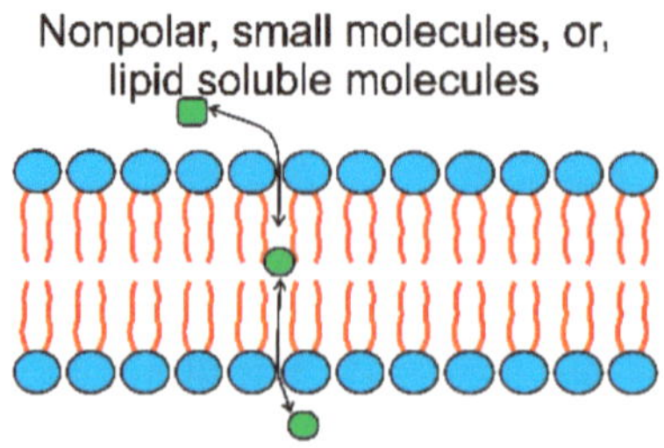

Figure 12.6: Diffusion across the phospholipids bilayer of the cell membrane

Since diffusion is a passive diffusion process, no metabolic energy (ATP) is involved and substances can only move down their concentration gradient. Lipid diffusion cannot be controlled by the cell, in the sense of being switched on or off.

The rate of diffusion across membranes depends on the following factors.

a) **Surface area of membrane:** the rate of diffusion is directly proportional to surface area of the membrane. The larger the surface area, the greater is the rate of diffusion.

b) **Difference in concentration across the membrane:** rate of diffusion is directly proportional to the concentration gradient. The larger the concentration gradient, the greater is the rate of diffusion.

c) **Thickness of membrane:** rate of diffusion is *inversely* proportional to the thickness of the membrane or the diffusion distance. Since cell membranes are extremely thin, the rate of diffusion will be high.

d) **Temperature:** rate of diffusion is directly proportional to the temperature as the kinetic energy of particles increase with temperature. The higher the temperature, the greater is the rate of diffusion.

e) **Size of particles:** Smaller and lighter particles diffuse faster, but it will also depend on the permeability of the membrane.

Facilitated Diffusion

Facilitated diffusion is the **net** movement of substances across a cell membrane, **through a trans-membrane protein molecule**, **without using ATP**. The particles ***move in both directions*** through the protein, but the rate of movement of particles from higher to lower concentration is greater than the movement in the opposite direction.

- The transport proteins tend to be **specific** for one type of molecule, so substances can only cross a membrane if it contains the appropriate protein.

- As the name suggests, this is a passive diffusion process, so **no metabolic energy (ATP)** is involved and substances can move down their concentration gradient by using **kinetic energy alone**.

- Some ions (Na^+, Ca^{2+}, K^+, Cl^-, HCO_3^-) and polar molecules (Glucose, amino acids) can diffuse through special transport proteins called **channel proteins or carrier proteins**. Diffusion can occur through the channel in either direction. Since diffusion would not be possible without these proteins the process is called facilitated diffusion.

- The rate of facilitated diffusion can be limited by the number of transport proteins per unit area of the membrane. Facilitated diffusion is a term which is specific to biology as it can only occur in cells.

There are two kinds of transport proteins that may be involved in facilitated diffusion:

Channel Proteins

Channel proteins are integral lipoproteins which form a water-filled pore or channel in the membrane, as shown in figure 12.7.

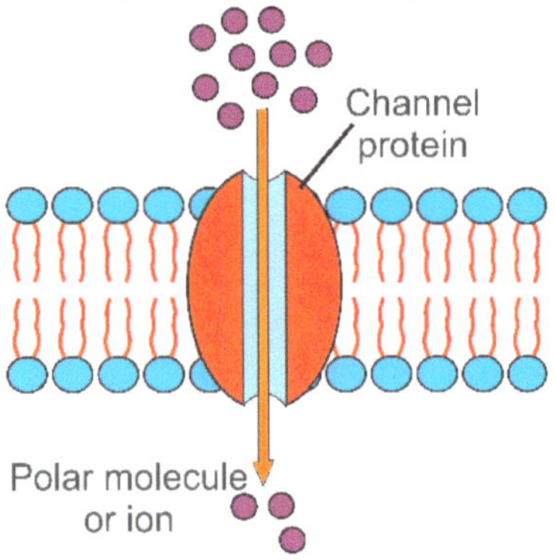

Figure 12.7: Channel protein

- The channel proteins allow ions to cross from one side of the membrane to the other.
- Channel proteins are ion-selective and may be gated (open or closed) to regulate the passage of ions in response to certain stimuli.
- Channel proteins only move molecules along a concentration gradient (i.e. are not used in active transport).
- Channel proteins have a much faster rate of transport than carrier proteins.

Carrier Proteins

Carrier proteins are integral glycoproteins which bind to a solute and undergo a *conformational change* to translocate the solute across the membrane.

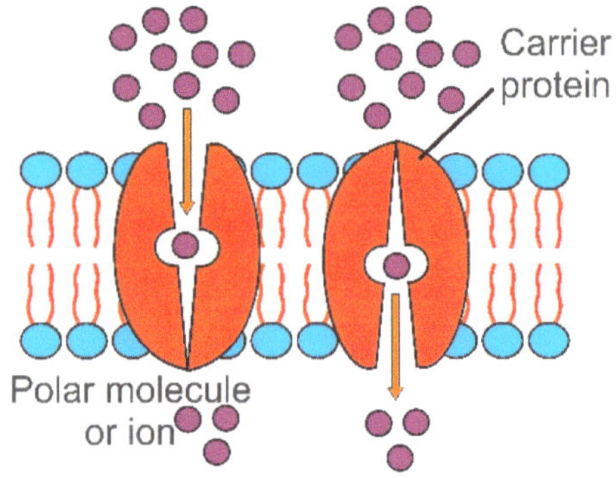

Figure 12.8: Carrier protein

- Carrier Proteins have a binding site for a specific solute and constantly flip between two states so that the site is alternately open to opposite sides of the membrane.
- In facilitated diffusion, there will be **net** movement of the substance from a high concentration to a low concentration, as shown in figure 12.8.
- Carrier proteins may move molecules against concentration gradients in the presence of ATP (i.e. are used in active transport)
- Carrier proteins have a much slower rate of transport than channel proteins (by an order of ~1,000 molecules per second)

Active Transport

Active transport is the transport of molecules or ions across the cell membrane through **carrier proteins** by using **energy from respiration (ATP).** It transports substances against the concentration gradient.

How active transport works

- The molecule or ion combines with a specific carrier protein in the cell surface membrane.
- ATP transfers a phosphate group to the carrier protein on the **inside of the membrane**. This causes the carrier protein to undergo a change of shape which causes the molecule or ion to move across the membrane.
- The molecule or ion is then released and the protein changes back to its original shape.
- The proteins are highly specific, so there is a different protein pump for each type of molecule or ion to be transported. The protein pumps are also ATPase enzymes, since they catalyse the splitting of ATP into ADP + phosphate (Pi), and use the energy released to change shape and pump the molecule. Pumping is therefore an active process and is the commonly used transport mechanism that can transport substances up their concentration gradient.

- The process of active transport is illustrated in figure 12.9.

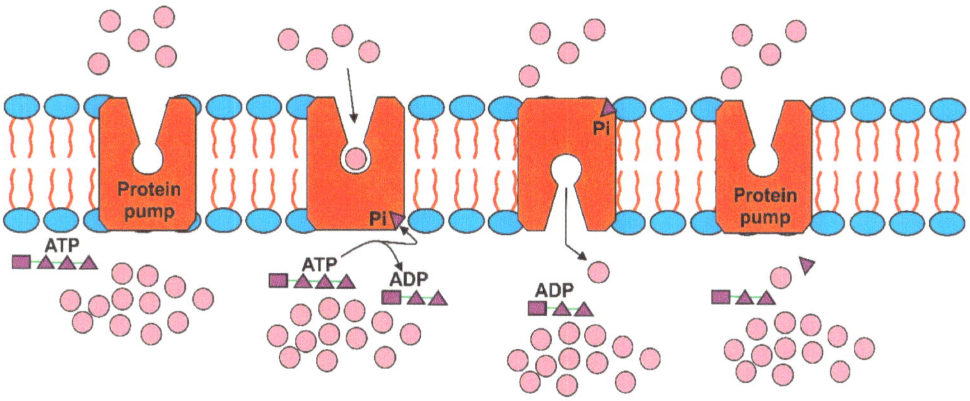

Figure 12.9: Active transport

Due to energy needed for this process, the cells involved tend to contain **more mitochondria** and have a high rate of respiration. Their ability to carry out active uptake is affected by temperature, oxygen concentration and the presence of respiratory poisons like Cyanide. **Factors which adversely affect the rate of respiration and ATP production will slow down active transport**.

Some processes involving active transport are nerve impulse transmission, muscle contraction, absorption of amino acids in ileum, absorption of ions by root hair cells of plants, selective re-absorption in kidney.

Cytosis – bulk transport involving Vesicles

The processes described so far only apply to small molecules. Large molecules, such as proteins and polysaccharides and even whole cells are moved in and out of cells by using vesicles. ATP is used in the movement of vesicles within the cells.

Endocytosis

Endocytosis is the transport of materials **into** a cell by forming vesicles. Materials are enclosed by a fold of the cell membrane, which then deepens and pinches of from the cell membrane to form a closed vesicle, as shown in figure 12.10.

If the materials are large **solids,** such as a white blood cell ingesting a bacterial cell or amoeba engulfing bacteria, the process is known as **phagocytosis (cell eating).** These vesicles are relatively large and are called as phagosomes. When the materials being taken into vesicles are **liquids or solutions,** such as a protein molecule and the process is known as **pinocytosis (cell drinking).** These vesicles are relatively smaller.

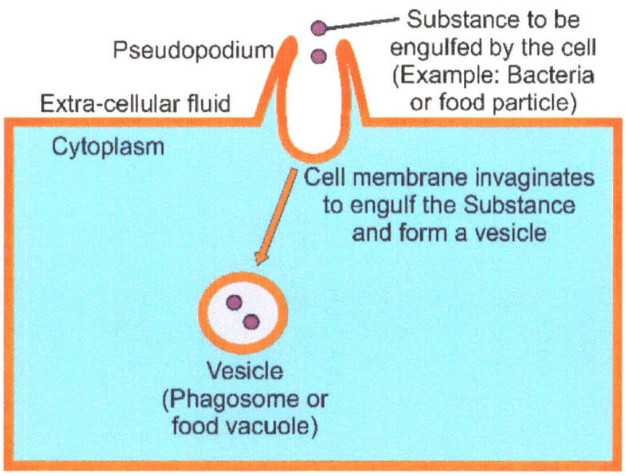

Figure 12.10: Endocytosis

Exocytosis

Exocytosis is the transport of materials **out of a cell by using vesicles**. It is the exact reverse of endocytosis. Materials to be secreted must first be enclosed in a membrane vesicle, usually from the Rough Endoplasmic Reticulum and Golgi bodies. The vesicles move towards the cell membrane and fuse with the cell membrane releasing their contents into the extracellular fluid, as shown in figure 12.11.

Hormones and digestive enzymes are secreted by exocytosis from the secretory cells of the intestine and endocrine glands.

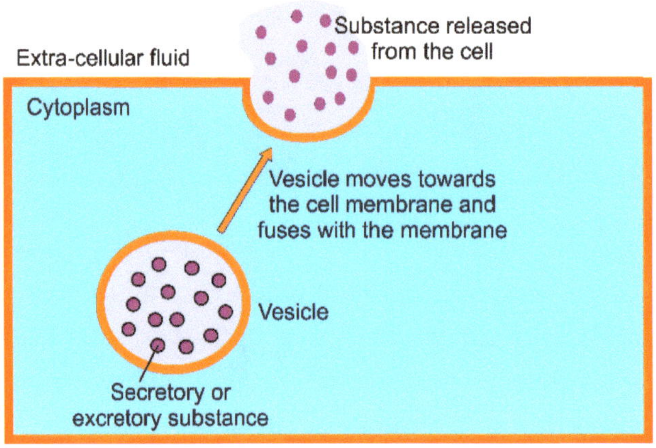

Figure 12.11: Exocytosis

Self-assessment (Chapter 12)

Step 1: *Visit www.teachingtube.com*

Step 2: *Sign up for the free course "Edexcel AS and AL Bio assessments"*

Step 3: *Attempt the chapter Quiz and get instant results.*

Step 4: *Download the assessment worksheet and marking scheme too.*

CHAPTER THIRTEEN
PROTEINS

Edexcel IAL from 2018 Specification 2.6: (i) know the basic structure of an amino acid (*Structures of specific amino acids are not required*).
(ii) understand the formation of polypeptides and proteins (amino acid monomers linked by condensation reactions to form peptide bonds)
(iii) understand the significance of a protein's primary structure in determining its secondary structure, three-dimensional structure and properties (globular and fibrous proteins and the types of bonds involved in its three-dimensional structure)
(iv) know the molecular structure of a globular protein and a fibrous protein and understand how their structures relate to their functions (including haemoglobin and collagen)

Polypeptides and proteins are chains of amino acids linked to each other by peptide bonds. All the proteins that are needed by humans can be made by the sequential arrangement of 20 amino acids.

Structure of an amino acid

The general structure of an amino acid is shown in figure 13.1. The amino (- NH_2) group and carboxylic acid (- COOH) groups are involved in the formation of **peptide bonds,** which hold amino acids together to form a polypeptide chain. The Residual (R) groups are involved in hydrogen bond, ionic bond, and covalent bond formation, which cause the polypeptide chain to fold. Each amino acid differs from another amino acid in the composition of the Residual group. There are 20 different amino acids that make up all the proteins in living organisms. Each of the 20 amino acids has a different residual group.

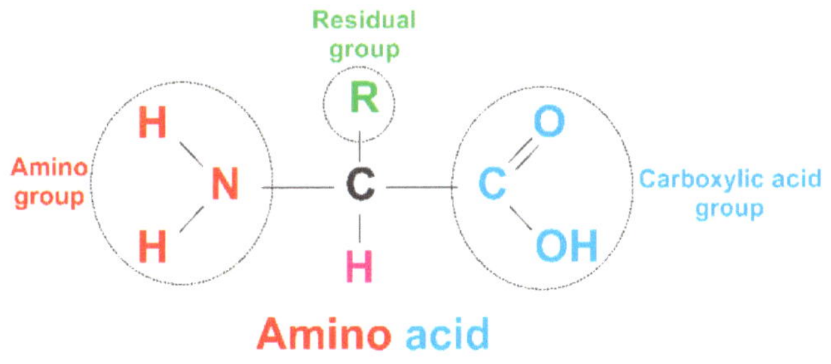

Fig. 13.1: Amino acid structure

Amino acids are linked to each other by the formation of **a peptide bond,** as shown in figure 13.2. The bond forms by a **condensation reaction** between the acid (- COOH) group of one amino acid and the amino (- NH_2) group of another amino acid. Two amino acids linked by a peptide bond is called a **dipeptide**, while many amino acids linked by peptide bonds forms a **polypeptide**.

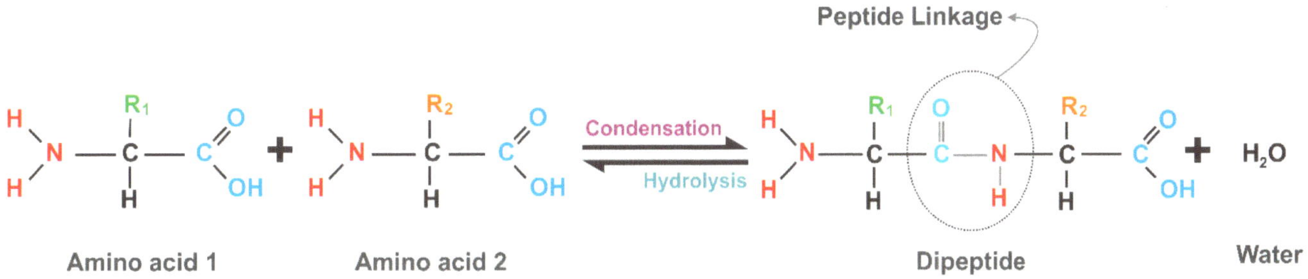

Figure 13.2: Condensation and hydrolysis

Condensation is the joining of two amino acids by the removal of a water molecule to form a peptide bond.

Hydrolysis is the splitting of a dipeptide or polypeptide by the addition of water molecules to break peptide bonds. Both condensation and hydrolysis are brought about by the action of enzymes.

Primary structure of a protein

The primary structure is the **sequence of amino acids** in a polypeptide chain. This sequence is determined by the genetic code on DNA. Finding the primary structure of a protein is called **protein sequencing**. The primary structure determines the secondary and tertiary structure of a protein.

For example, polypeptide chain A and polypeptide chain B differ in the type, number and sequence of amino acids. The eventual shape, properties and function of both polypeptide chains is going to be different, as the primary structure is different.

Polypetide chain A: 

Polypetide chain B: 

However, if two polypeptides have the same primary structure then their secondary and tertiary structure will be similar because they will form similar bonds between the amino acids.

Secondary structure of a protein

The secondary structure refers to the folding of the polypeptide chain (primary structure) into helices and pleated sheets. The secondary structure is held together **by hydrogen bonds between the carboxyl groups residues (-C=O) and the amino group residues (-NH) in the polypeptide backbone**.

Two secondary structures, the Alpha (α) helix and the Beta pleated (β) pleated sheet are described below.

The alpha(α) helix

The polypeptide chain is wound to form a helix. It is held together by hydrogen bonds running parallel with the long helical axis. There are so many hydrogen bonds that this is a very stable and strong structure. Helices are common structures throughout biology.

Example: Keratin in hair, Nails and Skin.

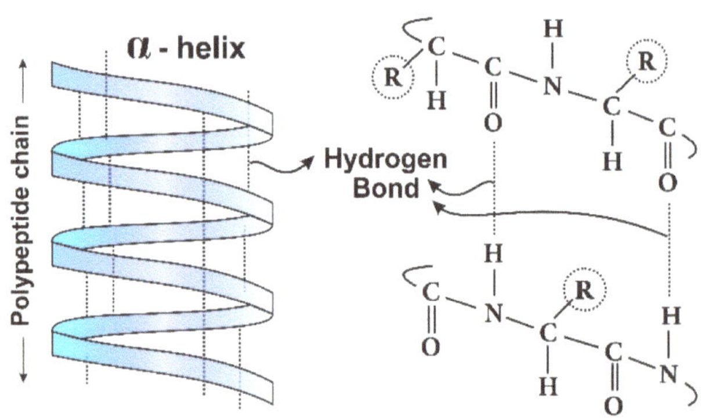

Figure 13.3 - Alpha(α) helix

The β pleated – sheet

The polypeptide chain zig-zags back and forward forming a sheet. Once again it is held together by hydrogen bonds.

Example: Fibroin, in silk.

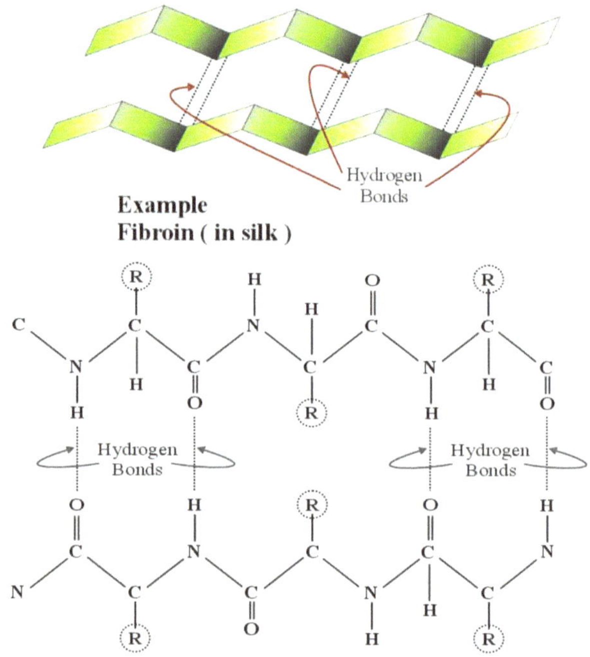

Figure 13.4 – Beta (β) pleated sheet

Tertiary structure of a protein

The tertiary structure of a protein is the complex three - dimensional **globular** shape the polypeptide chain takes when the polypeptide chain twists and folds around itself. The tertiary structure is maintained by Hydrogen bonds, disulphide bridges [covalent bonds] and ionic bonds between the **Residual groups of amino acids**. Hydrophobic interactions also help to maintain the shape of globular proteins, like enzymes and trans-membrane proteins.

The specific three-dimensional shape [secondary, tertiary, quaternary structure] of a protein is maintained by three types of chemical bonds between the Residual groups of amino acids

1. **Hydrogen bonds:** form between some Hydrogen atoms [which bear a slight positive charge] and oxygen and nitrogen atoms [which bear a slight negative charge]. Although these bonds are weak, the large number of bonds provide a considerable force to maintain the three-dimensional shape.
2. **Ionic bonds:** form between carboxyl [COOH] groups and amino [NH_2] groups found in the Residual chains. They are stronger than H bonds, but can be broken by changes in pH and high temperature
3. **Disulphide bonds:** Some amino acids, like cysteine and methionine contain sulphur atoms in the Residual groups. Disulphide bonds form between sulphur atoms of amino acids that are close together. These bonds are strong and contribute to the strength of structural proteins like collagen. They are also useful in linking the two polypeptide chains of insulin together.

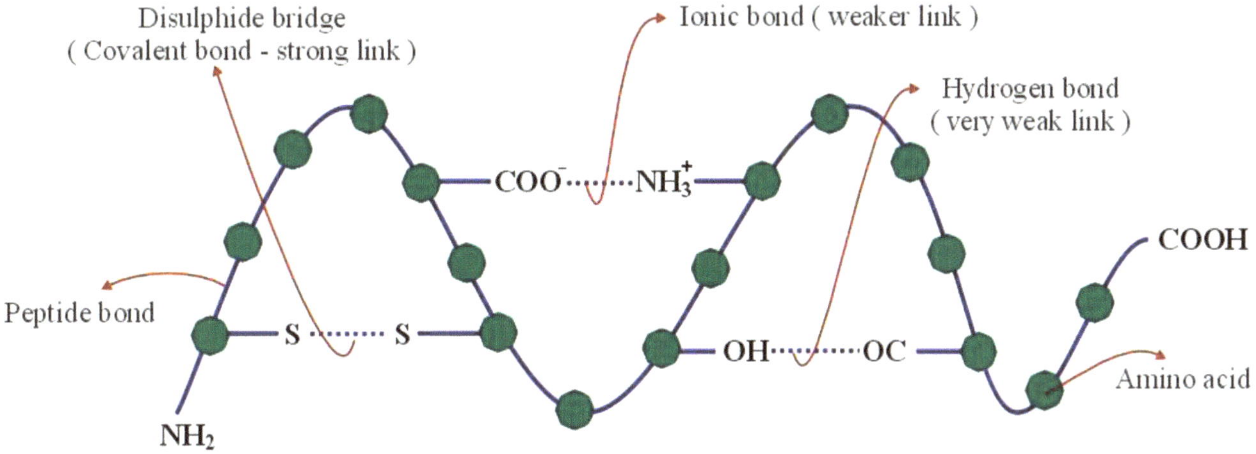

Figure 13.5 – Bonding between the residual groups

Hydrophilic and hydrophobic interactions

Hydrophilic and hydrophobic interactions also help to maintain the shape of globular proteins [tertiary structure] in aqueous solutions. The hydrophobic [water hating] regions of the polypeptide chain face away from water by folding inwards. The hydrophilic regions of the chain remain on the surface of the globular structure.

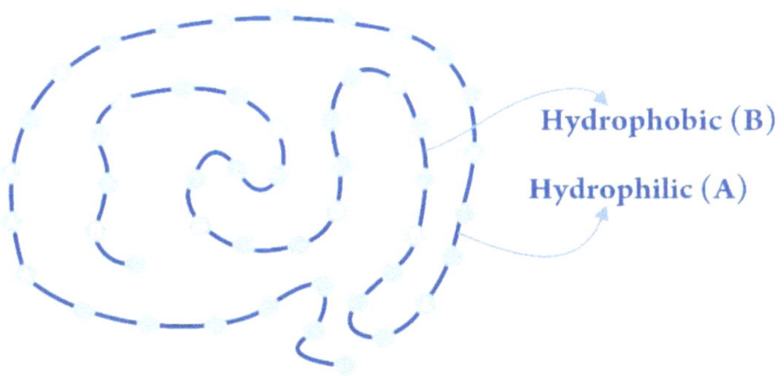

Figure 13.6 – Hydrophilic and hydrophobic interactions

Quaternary structure of a protein

Quaternary structure is the linking together of two or more polypeptide chains.

An example of a protein with a quaternary structure is haemoglobin (O_2 carrying molecule in red blood cells).

Haemoglobin Structure

Haemoglobin is composed of four polypeptide chains (two alpha chains and two beta chains). The alpha globin chains have 141 amino acids and each beta globin chain has 146 amino acids. The sequence of amino acids in these chains determines the shape and properties of the quaternary structure of haemoglobin. It is also composed of iron-containing haem-groups (prosthetic groups responsible for binding oxygen).

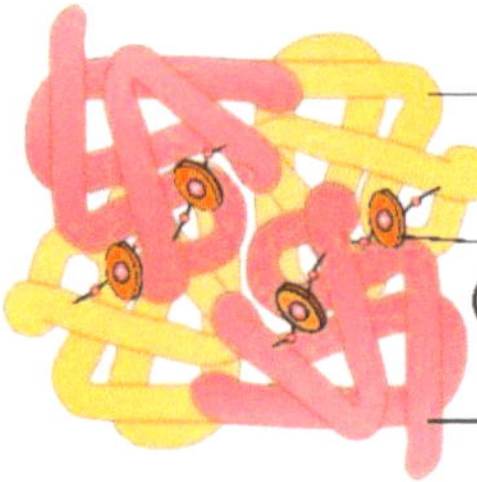

Figure 13.7 – Haemoglobin Structure

Collagen Structure

Collagen is a fibrous protein. It has a quaternary structure, since it is made up of three polypeptide chains. Each polypeptide chain is twisted to form a helix. The three polypeptide helices wind around each other like a rope with three strands. Hydrogen bonds hold the three strands in place. This makes collagen very stable, insoluble, flexible, but inelastic. Collagen is found mainly in bones and tendons.

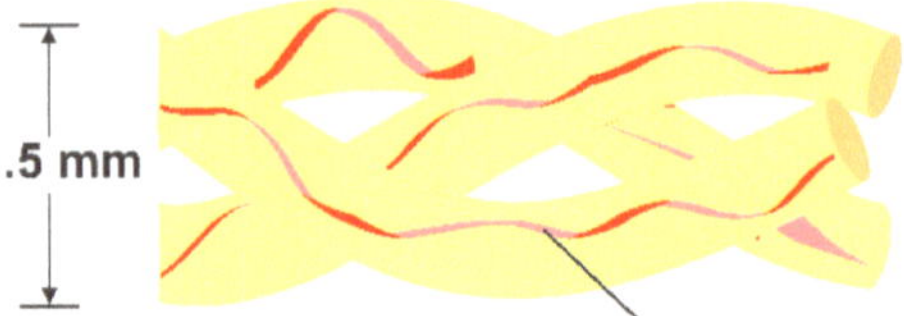

Figure 13.8 – Collagen Structure

Differences between Fibrous and Globular Proteins

	Fibrous proteins	Globular proteins
Shape	Long and thread-like	Rounded / spherical
Purpose	Structural proteins	Functional proteins
Durability	Less sensitive to changes in pH, temperature, etc.	More sensitive to changes in pH, temperature, etc.
Examples	Collagen, myosin, fibrin, actin, keratin, elastin	Enzymes, haemoglobin, insulin, immunoglobin
Solubility	Insoluble in water	Soluble in water

Edexcel IAL from 2018 Specification: RECOMMENDED ADDITIONAL PRACTICAL
Use a semi-quantitative method to estimate protein concentration using biuret reagent and colour standards.

Semi-quantitative test for estimation of protein concentration

- *Prepare a range of protein (albumin) solutions with concentrations of 0%, 0.1%, 0.5%, 1.0 %, and 5.0%.*
- *Equal volumes of the 0.1% protein solution and Biuret's reagent solution is pipetted into a test tube. The procedure is repeated with all the other solutions.*
- *The colour of the solutions will change from blue to purple or lilac, as shown in figure 13.9. The shade of the colour depends on the concentration of the protein in the solution. These are known as colour standards.*

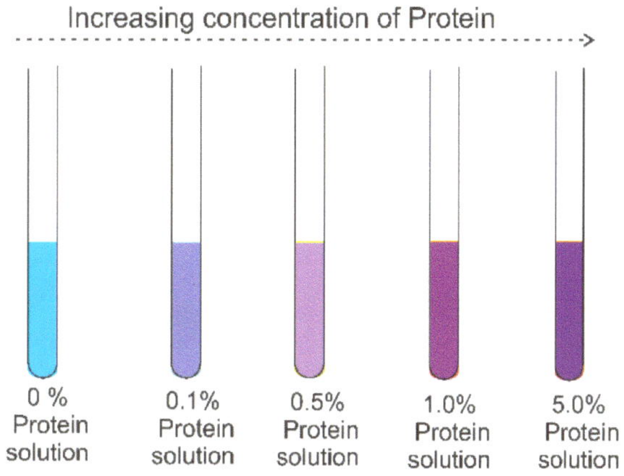

Figure 13.9: Biuret's test for proteins (Semi-quantitative)

- *Repeat the procedure with a protein solution of unknown concentration, or, with a food extract. Compare the colour of this solution with the colour standards to estimate the protein concentration in the unknown solution.*

Self-assessment (Chapter 13)

Step 1: *Visit www.teachingtube.com*
Step 2: Sign up for the free course "Edexcel AS and AL Bio assessments"
Step 3: Attempt the chapter Quiz and get instant results.
Step 4: Download the assessment worksheet and marking scheme too.

CHAPTER FOURTEEN
ENZYMES

Edexcel IAL from 2018 Specification 2.7: (i) understand the mechanism of action and the specificity of enzymes in terms of their three-dimensional structure
(ii) understand that enzymes are biological catalysts that reduce activation energy
(iii) know that there are intracellular enzymes catalysing reactions inside cells and extracellular enzymes catalysing reactions outside cells

Enzyme structure

All enzymes are **globular proteins** - the polypeptide chain is folded into a spherical or globular shape. Hydrogen bonds, ionic bonds, disulphide bridges and hydrophobic interactions (between R groups of amino acids) maintain the specific three-dimensional shape of the enzyme, as shown in figure 14.1.

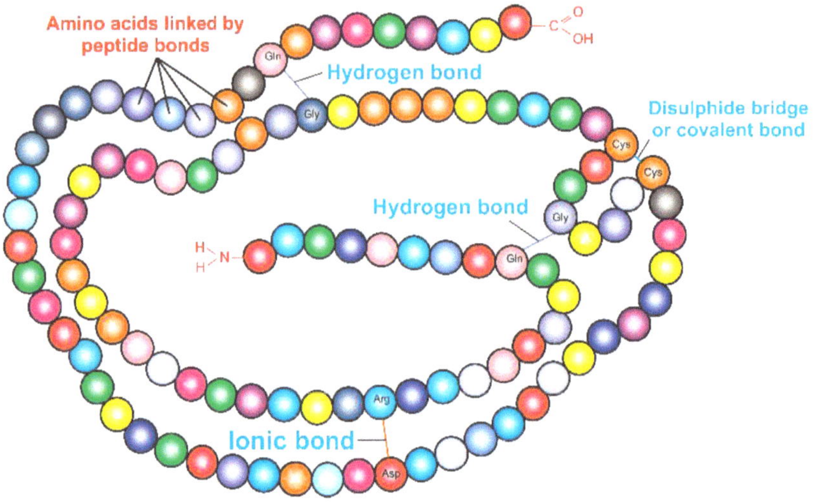

Figure 14.1: Structure of an enzyme

Enzyme specificity – Lock and key hypothesis

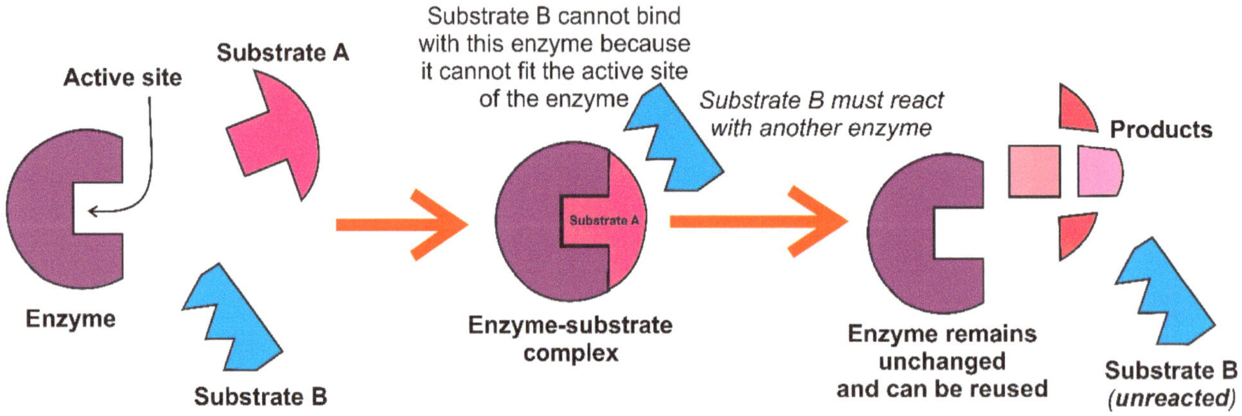

Figure 14.2: Lock and key hypothesis

This specific tertiary shape is very essential for the functioning of enzymes. The part of the enzyme which reacts with the substrate is called the **active site**. The shape of the active site differs from one enzyme to another. This makes the enzyme react only with a **specific substrate**, which fits the active site, as in shown in figure 14.2.

Activation energy

Activation energy is the minimum energy that the reactant molecules must possess to start a reaction. Thus, to start a reaction, energy must be supplied to the substrate molecules. This energy is called the activation energy. Enzymes lower the activation energy and provide an alternate, lower energy pathway for the reaction to proceed, as shown in figure 14.3. Thus, the rate of reaction speeds up. So, enzymes are referred to as **biological catalysts**.

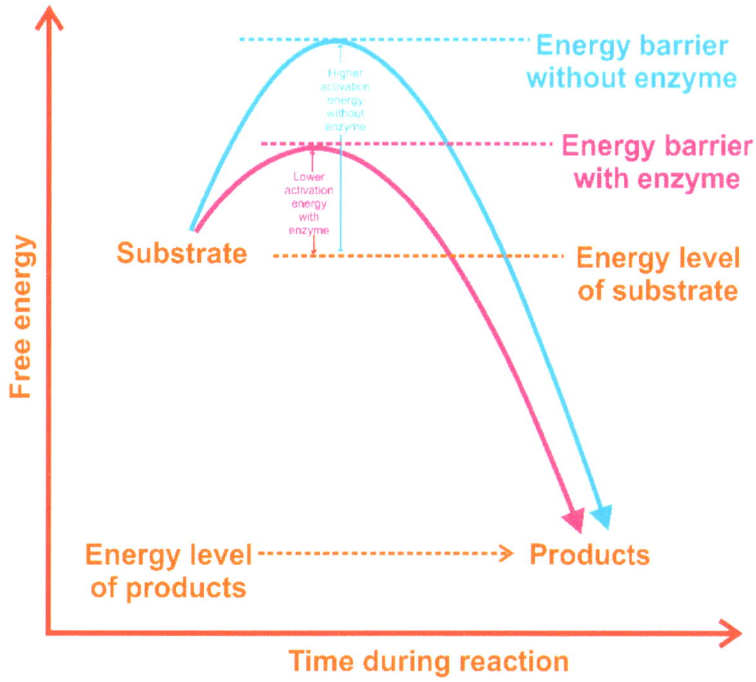

Figure 14.3: Enzymes lower activation energy

Induced fit hypothesis – lowering activation energy

The lock and key hypothesis explained the specificity of enzyme action; however it could not explain how enzymes lower the activation energy. The induced fit model of enzyme activity explains how enzymes reduce activation energy, as shown in figure 14.4.

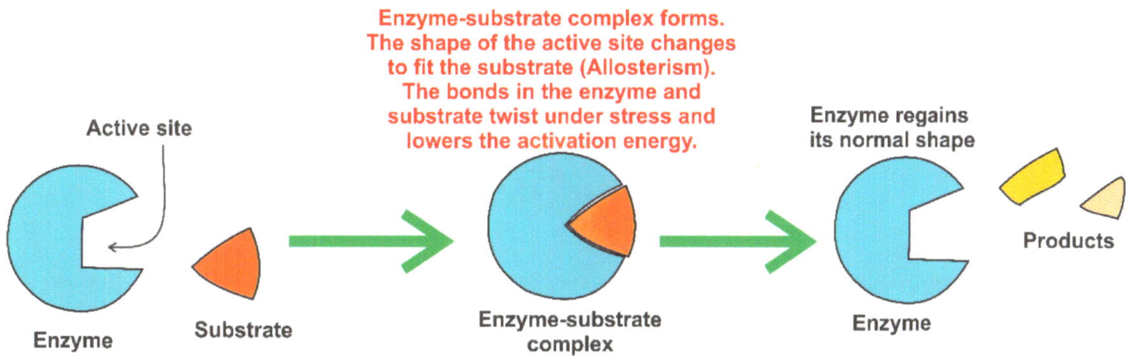

Figure 14.4: Induced fit hypothesis

Other factors that could affect shape of enzymes

Temperature

Temperature can influence the shape of enzymes. As temperature increases up to the optimum, the rate of enzyme activity also increases. This is because enzyme and substrate molecules gain more kinetic energy and the collisions between active sites of enzymes and substrate molecule become *more frequent*. The *rate* of enzymes substrate complex formation increases so enzymes activity speeds up.

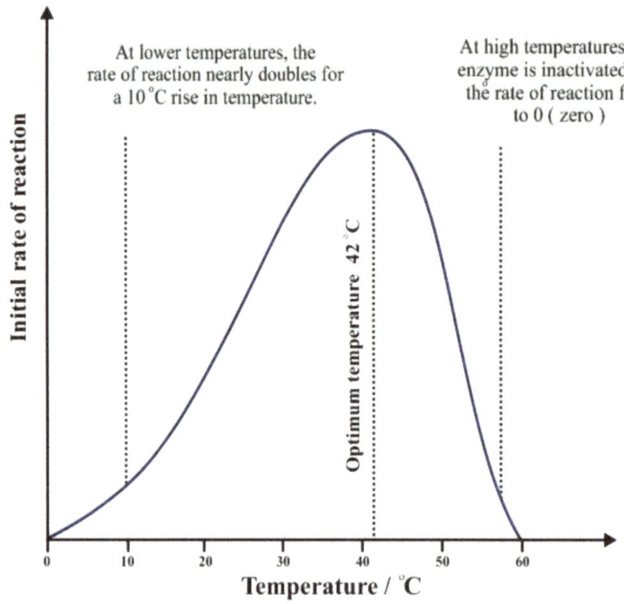

Figure 14.5: Effect of temperature on the rate of enzyme action

Beyond the optimum temperature, the rate of enzyme activity decreases because the high temperature causes the enzyme molecule to lose its specific 3D shape, due to breaking of Hydrogen bonds, ionic and covalent bonds. The active sites cannot bind with substrate, so enzyme substrate complexes cannot form. The optimum temperature of enzymes may differ depending upon the type of bonds that maintain the tertiary structure. Enzymes with many hydrogen bonds will denature easily as these bonds are relatively weak.

Non-active site directed inhibitors

These are **non- competitive** inhibitors which bind to the enzyme molecule at an **allosteric site** and change the shape of the active site. This will prevent formation of enzyme substrate complex formation, reducing the rate of enzyme action.

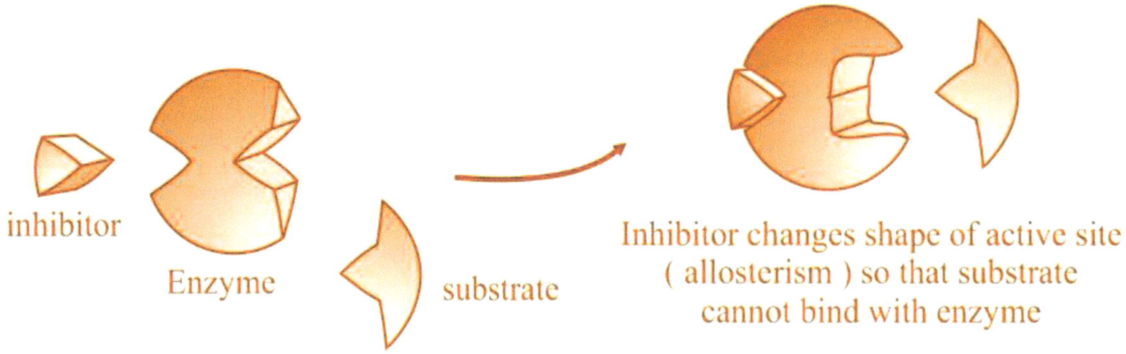

Figure 14.6: Effect of inhibitors on the rate of enzyme action

Effect of pH on enzyme activity

Enzymes are very sensitive to changes in pH. Every enzyme works best within a very narrow range of pH. If the pH changes above or below the optimum, then rate of reaction will decrease.

This is because at the optimum pH the H^+ ions / OH^- ionic balance (pH) of the solution is just right to maintain the specific 3D tertiary structure of the enzyme.

If pH changes then the change in H^+ ion concentration will disrupt the ionic bonds which maintain the tertiary structure of the enzyme. This will cause the enzyme to change its shape so that the active site cannot bind with substrate molecules. Enzyme substrate complexes will form at a slower rate, so rate of enzyme activity decreases.

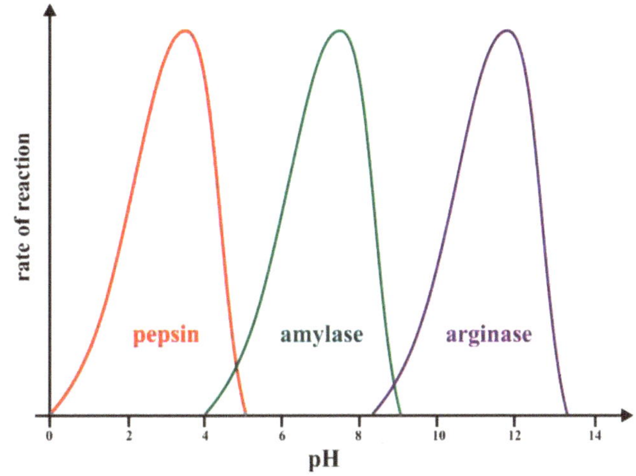

Figure 14.7: Effect of pH on the rate of enzyme action

Roles of enzymes – controlling metabolic pathways

A metabolic pathway is a sequence of enzyme-controlled reactions, where every step in the pathway is controlled by a specific enzyme. These reactions may occur within cells (intracellular) or outside the cells (extracellular).

Digestion of starch is an example of extracellular metabolism.

Starch $\xrightarrow{\text{Amylase(hydrolase)}}$ **Maltose** $\xrightarrow{\text{Maltase(hydrolase)}}$ **Glucose**

Starch is the primary substrate, Maltose is an intermediate compound and glucose is the final product. Every reaction in the pathway is controlled by a specific enzyme. This enables the cell to have more control over metabolic reactions, by adjusting the concentration of enzymes to slow down or speed up specific reactions in the body. The reaction can be stopped by shutting down the secretion of the enzyme.

Glycolysis is an example of an intracellular metabolic pathway that occurs during respiration in the cytoplasm of cells. The *simplified* metabolic pathway below shows some of the enzymes involved in glycolysis.

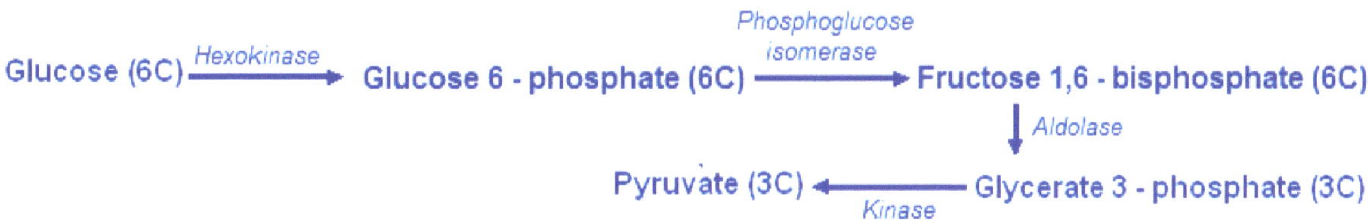

Edexcel IAL from 2018 Specification 2.8: CORE PRACTICAL 4
Investigate the effect of temperature, pH, enzyme concentration and substrate concentration on the initial rate of enzyme-catalysed reactions.

Step one: Preparation of Catalase solution (Enzyme solution)

Weigh 0.1 mole of catalase powder by using an electronic balance and a weighing boat.

Add distilled water to the catalase powder until the volume of the solution reaches 1dm^3. This gives a catalase solution of 0.1 mol dm^{-3} concentration.

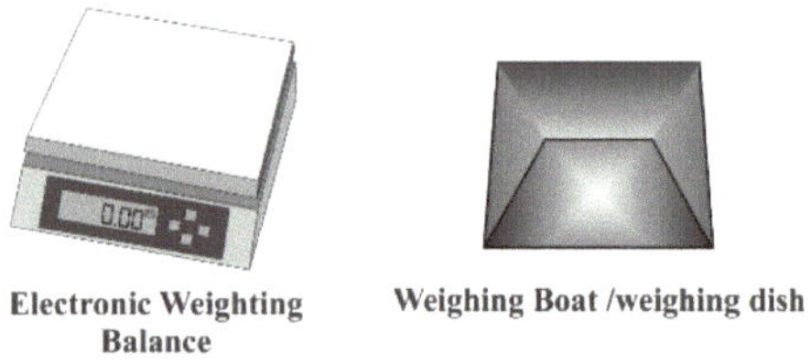

Figure 14.8: – Weighing the enzyme

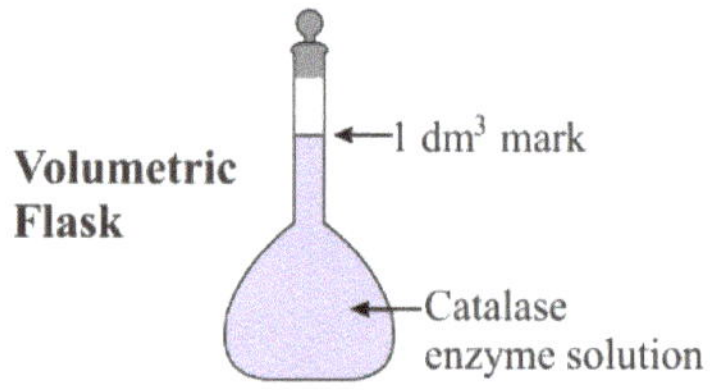

Figure 14.9: – Enzyme solution

Transfer 5cm^3 of this enzyme solution into a syringe and attach it to the apparatus shown in step two.

Note: the molar mass of catalase enzyme will be stated on the label of the container. A range of catalase concentrations can be prepared by dissolving a specific number of moles of the enzyme into 1 dm^{-3} distilled water.

Step Two: Preparing the substrate solution (Hydrogen peroxide solution)

A standard hydrogen peroxide solution of 3% concentration (1 M) will be used for all the trials. The volume of hydrogen peroxide solution should also be kept constant at 10 cm^3.

Transfer 10 cm^3 of hydrogen peroxide solution into a conical flask and set up the apparatus as shown below.

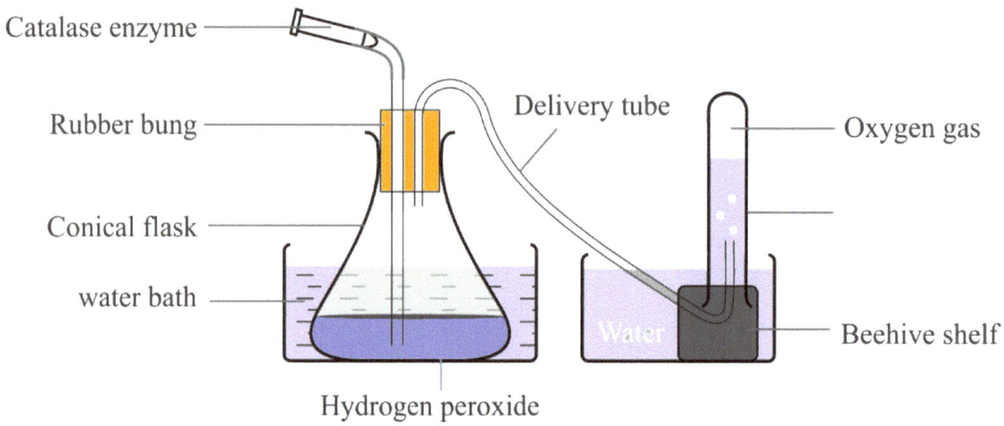

Figure 14.10: – Measuring the volume of oxygen evolved

Step Three: Measuring the initial rate of reaction

Put the bung securely in the flask – twist and push carefully.

Half-fill the trough, bowl or sink with water. Fill the 50 cm^3 measuring cylinder with water. Invert it over the trough of water, with the open end under the surface of the water in the bowl and with the end of the rubber tubing in the measuring cylinder. Clamp in place.

Attach the syringe with the enzyme in place in the bung of the flask but do not push the plunger straight away.

Check the rubber tube is safely in the measuring cylinder. Push the plunger on the syringe and immediately start the stop clock.

After every 10 seconds, note the volume of oxygen in the measuring cylinder in a suitable table of results.

Time / seconds	Volume of oxygen evolved / cm^3			
	Trial one	Trial two	Trial three	Mean
0				
10				
20				
30				
40				
50				
60				

Plot the results on a time course graph as shown below.

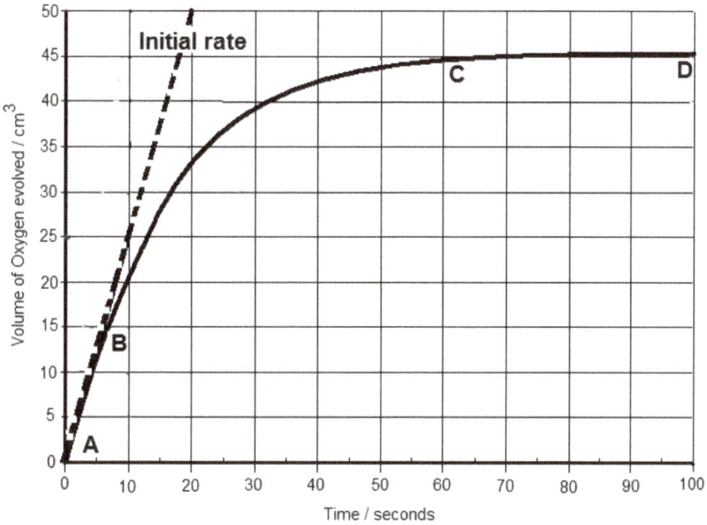

Figure 14.11: – Time course graph

Rate of reaction = Volume of oxygen formed / Time

A to B: Rate of oxygen production is constant. Enzyme concentration is a limiting factor. Substrate is in excess. **Gradient is equal to initial rate of reaction.**

B to C: Substrate concentration is becoming a limiting factor. Rate of oxygen production is slowing down. So, rate of reaction is decreasing.

C to D: Reaction is complete. Rate of oxygen production is zero. All the substrate has been used up.

Step Four: Repeating the experiment at varying enzyme concentrations

Empty and rinse the conical flask and measure another 10 cm^3 Hydrogen peroxide into it. Reassemble the apparatus, refill the measuring cylinder, and repeat steps two and three with another concentration of Catalase enzyme.

Chapter Fifteen

Nucleic Acids

Edexcel IAL from 2018 Specification 2.9: (i) know the basic structure of mononucleotides (deoxyribose or ribose linked to a phosphate and a base, including thymine, uracil, adenine, cytosine or guanine) and the structures of DNA and RNA (polynucleotides composed of mononucleotides linked by condensation reactions to form phosphodiester bonds)
(ii) know how complementary base pairing and the hydrogen bonding between two complementary strands are involved in the formation of the DNA double helix

Mononucleotide structure and formation

DNA and RNA belong to a group of compounds called **Nucleic acids**. The building blocks of nucleic acids are called **mononucleotides.** Ribonucleic acid (RNA) and deoxyribonucleic acid (DNA) are made up by linking of many mononucleotides, by condensation reactions.

A mononucleotide is made up of three parts, as shown in figure 15.1.

- An inorganic phosphate group;
- A pentose sugar - ribose in RNA and deoxyribose in DNA;
- A base which contains nitrogen - Adenine (A), Guanine (G), Cytosine (C), Uracil (U) **(only in RNA)**, Thymine (T) **(only in DNA)**

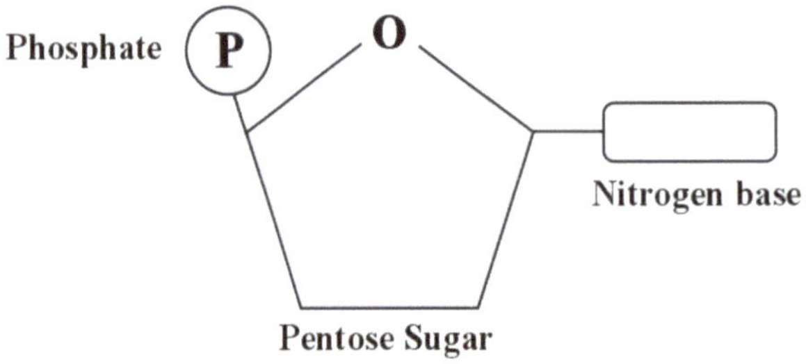

Figure 15.1: Structure of a mononucleotide

Condensation reactions join up the inorganic phosphate group and nitrogenous base to the sugar by removal of water molecules, as shown in figure 15.2.

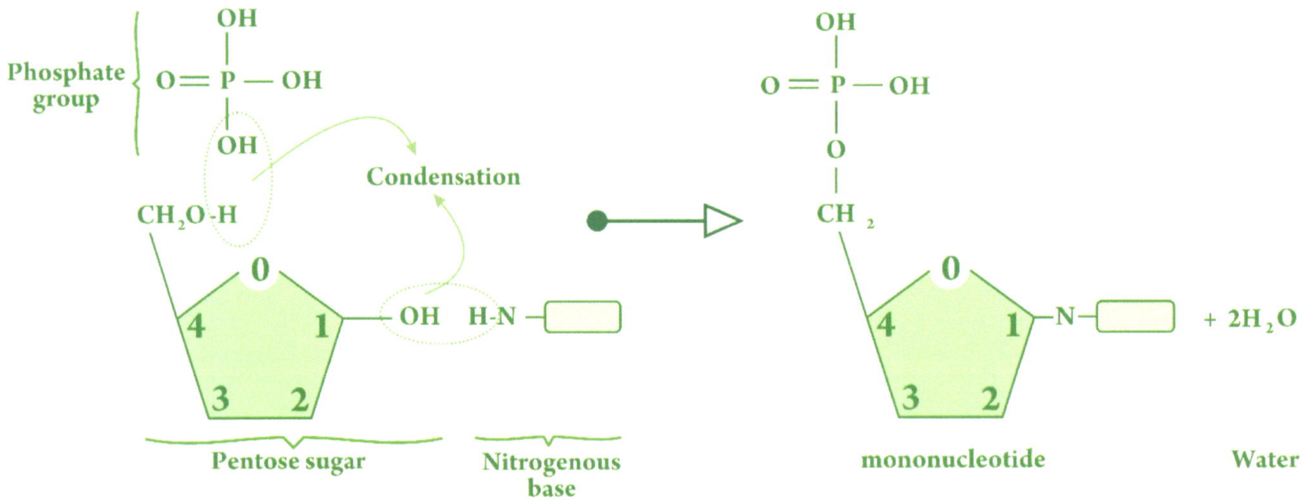

Figure 15.2: Formation of a mononucleotide by condensation reactions

Formation of a polynucleotide by condensation

Condensation reactions are involved in the formation of phospho-diester bonds. Carbon 5' of the pentose binds with carbon 3' of pentose on another nucleotide. A phosphate group links the two pentose molecules to each other. Many nucleotides can be linked by phospho-diester bonds to form a polynucleotide.

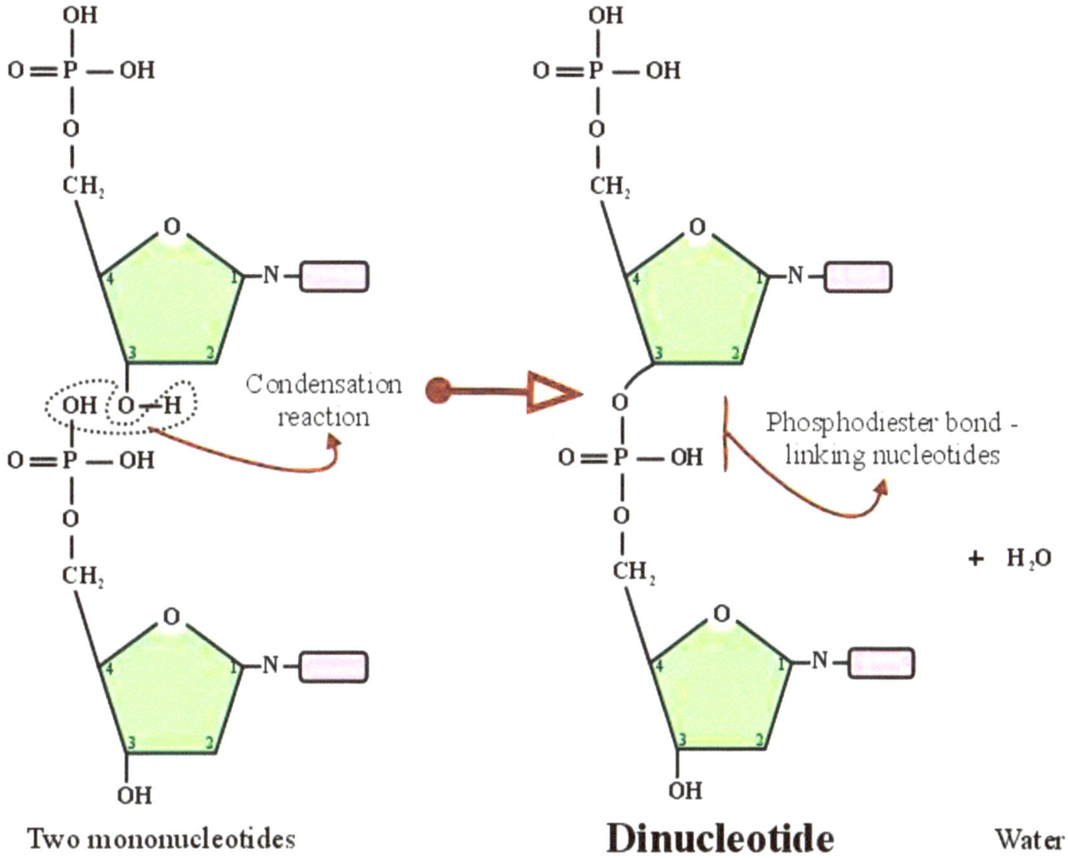

Figure 15.3: Formation of a dinucleotide by a condensation reaction

Structure of RNA

RNA is a single polynucleotide chain. The mononucleotides are linked to each other by phospho-diester bonds.

Each mononucleotide contains

- **a ribose sugar**,
- **a nitrogenous base (except thymine)** and
- **a phosphate group**.

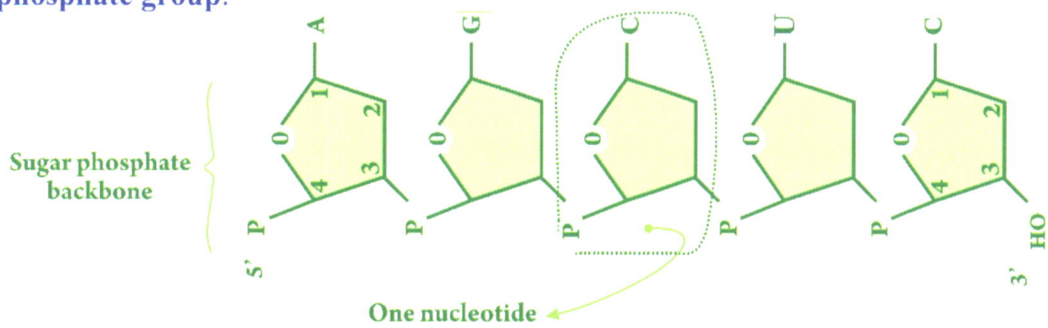

Figure 15.4: Structure of Ribo-Nucleic Acid (RNA)

Structure of DNA

DNA is a polymer of deoxyribo-nucleotides. Each mononucleotide in DNA consists of

- **a deoxyribose sugar,**
- **a nitrogenous base (except uracil)** and
- **a phosphate group.**

DNA consists of two polynucleotide strands which are anti parallel, one strand runs from 3′ to 5′ end and the other runs from 5′ to 3′ end. The two strands are linked to each other by **Hydrogen bonds between nitrogenous base pairs,** Adenine pairs with Thymine and Guanine pairs with cytosine. The strands are wound in to a **double helix**. The base pairs are 0.34 nm apart and there are 10 base pairs in one complete turn of the helix. Both strands are complementary to each other.

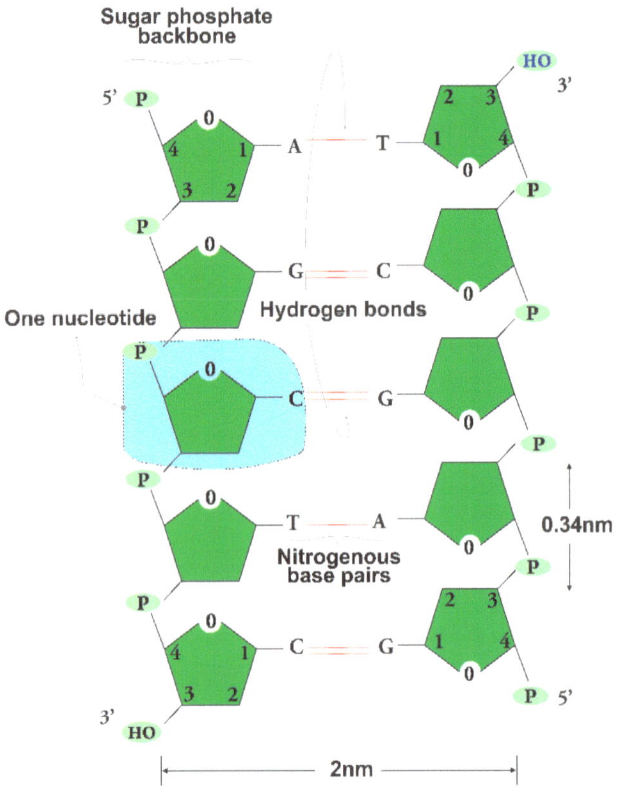

Figure 15.5: Complementary base pairing in DNA

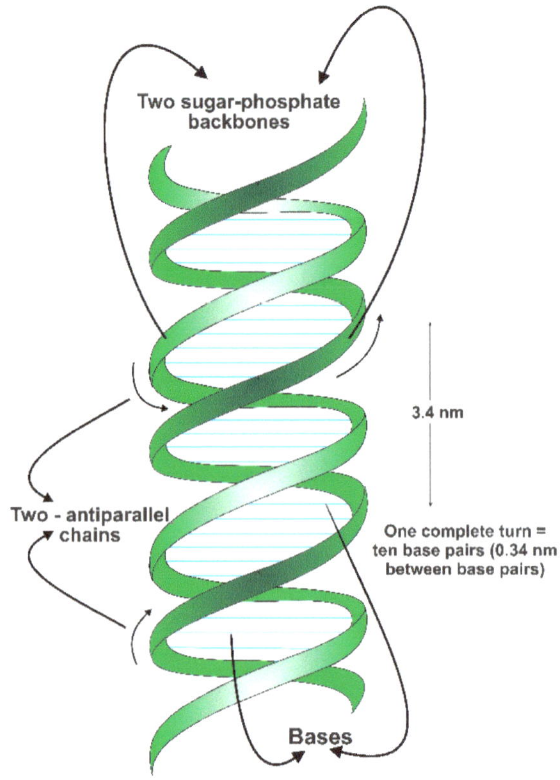

Figure 15.6: DNA double helix

Edexcel IAL from 2018 Specification 2.10: (i) understand the process of DNA replication, including the role of DNA polymerase
(ii) understand how Meselson and Stahl's classic experiment provided new data that supported the accepted theory of replication of DNA and refuted competing theories

DNA Replication

During semi conservative replication of DNA, each daughter DNA molecule consists of one complete old strand and one complete new strand of nucleotides, as shown in figure 15.7.

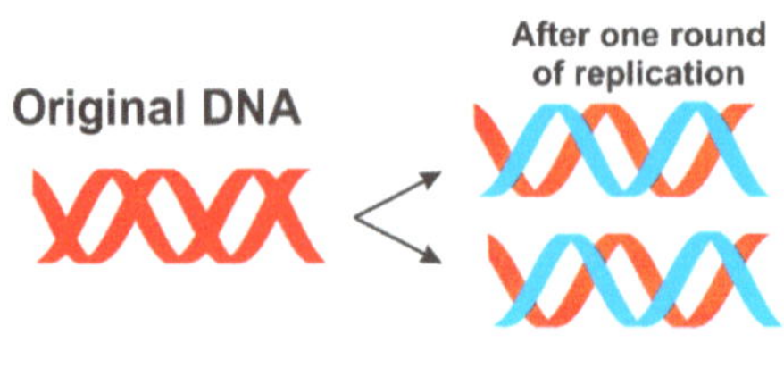

Figure 15.7: Semi-conservation replication of DNA

The process of semi conservative replication of DNA is illustrated below.

1. The two strands of DNA are held together by hydrogen bonds between the nitrogenous base pairs.

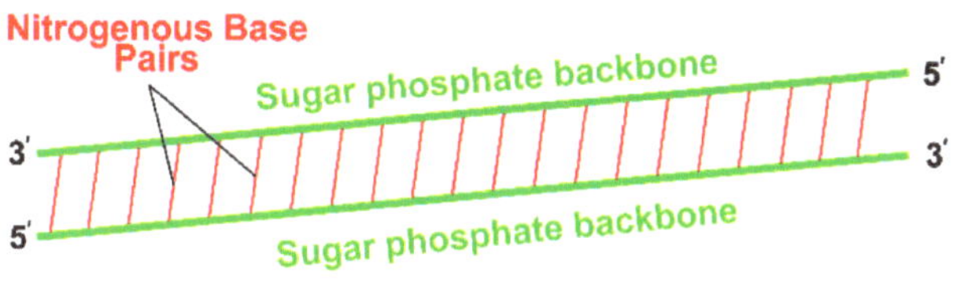

Figure 15.8: DNA double helix

2. The enzyme *Helicase* unwinds and unzips the DNA by breaking the hydrogen bonds between the nitrogenous base pairs. This exposes the nitrogenous bases on each strand.

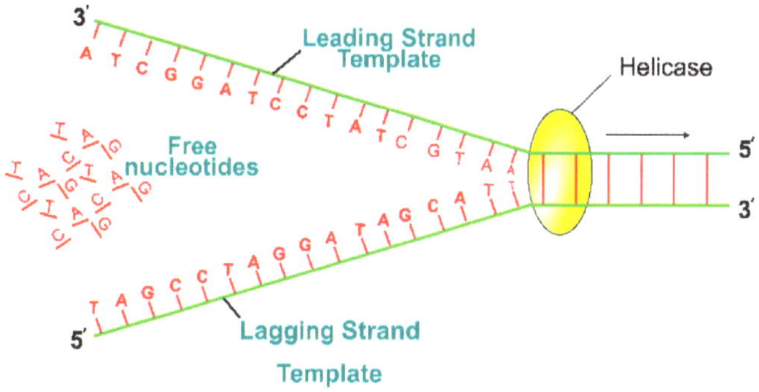

Figure 15.9: Helicase unzips DNA

3. Each DNA strand will serve as a **template** for free nucleotides to pair up with complementary nitrogenous bases, following the base pairing rule – Adenine with Thymine and Guanine with Cytosine.

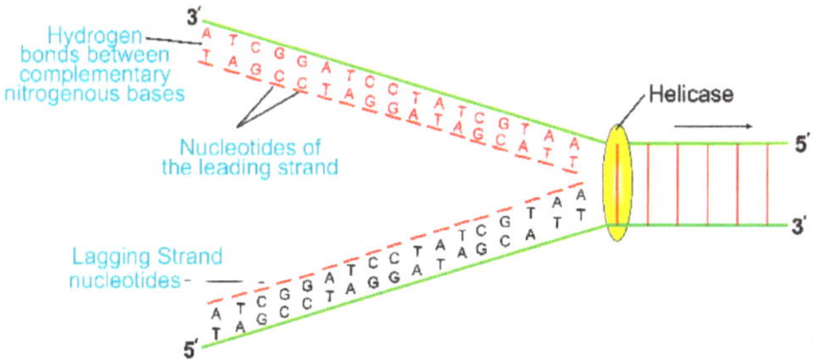

Figure 15.10: Complementary base pairing

4. The enzyme *DNA polymerase* can then join the nucleotides to each other by **phospho-diester bonds.**

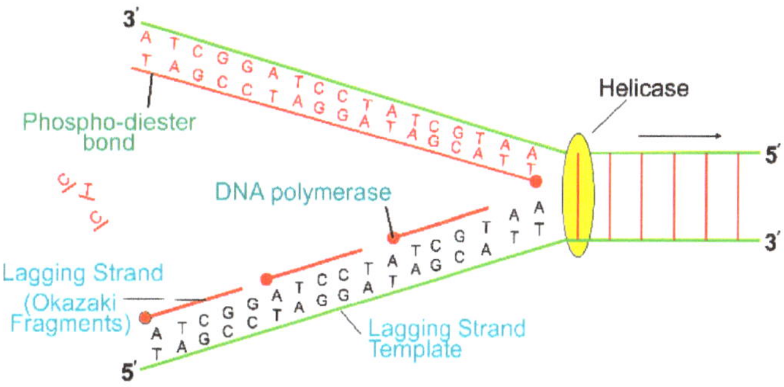

Figure 15.11: Condensation reactions by DNA polymerase

5. On the leading strand, the nucleotides are joined continuously but on the lagging strand, the new polynucleotide chain is formed in fragments *(Okazaki Fragments)*. These fragments are joined up by *DNA ligase.*

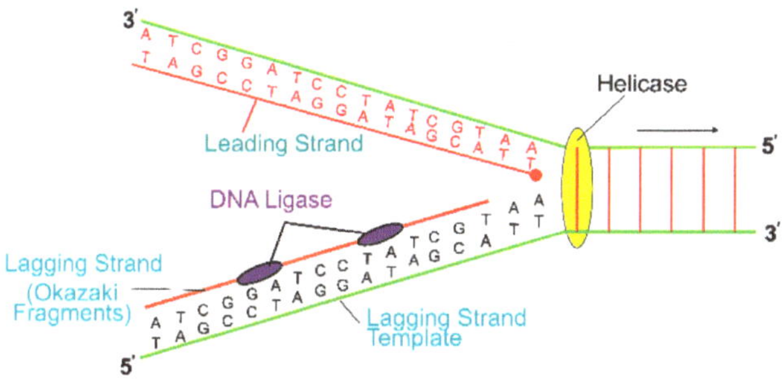

Figure 15.12: DNA ligase joins Okazaki fragments

6. Two identical molecules of daughter DNA are produced, but each daughter DNA molecule has one complete parent strand and one complete new strand of polynucleotides. This is called **semi-conservative replication.**

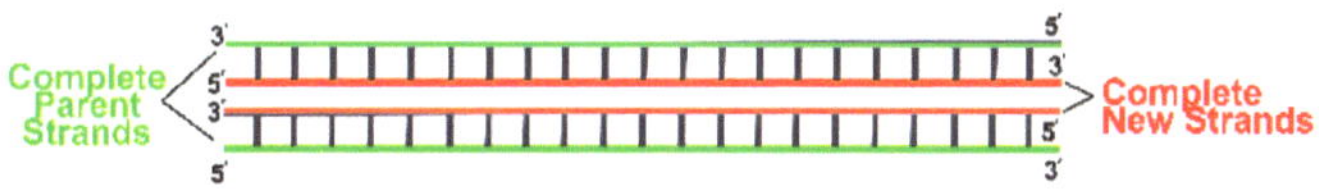

Figure 15.13: Semi-conservative replication

Competing theories

The mechanism of DNA replication was postulated in the form of three theories.

1. **Conservative replication: This theory** suggested that a "photocopy" of the original DNA was made, leaving the original DNA conserved.

2. **Semi-conservative replication: This theory** suggested that each new DNA molecule contains one complete new strand and one complete old strand.

3. **Dispersive replication: This theory** suggested that each daughter DNA molecule consisted of fragments of old and new nucleotide strands.

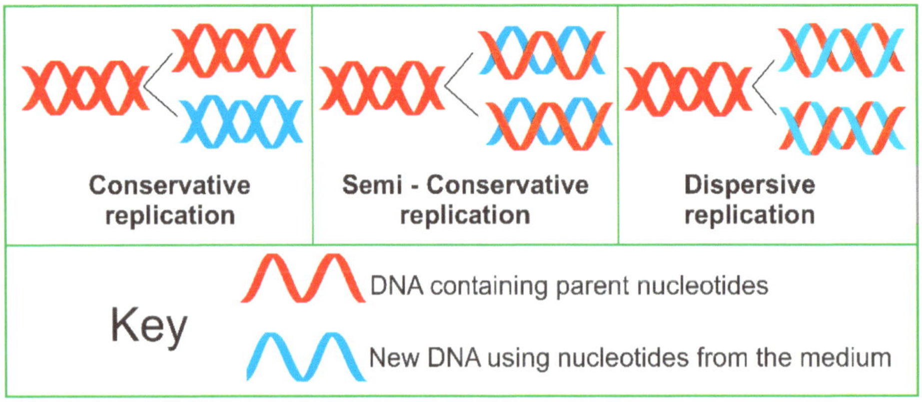

Figure 15.14: Competing theories of DNA replication

The proof that the semi-conservative method was the correct method came from an experiment performed by Meselson and Stahl using the bacterium *E. coli* together with the technique of density gradient centrifugation, which separates molecules based on their density.

Meselson and stahl density gradient experiment

Density gradient in caesium chloride solution

When a caesium chloride soultion is centrifuged, a density gradient is setup. The upper layers in the centrifuge tube have a minimum density. The density of the solution increases as we deeper into the solution. DNA is insoluble in caesium chloride and will settle into the soultion at the depth where the density od DNA matches the density of the solution. So, denser DNA will settle down deeper in the solution and lighter, less dense DNA will settle down just below the surface, where the density of the solution is lower.

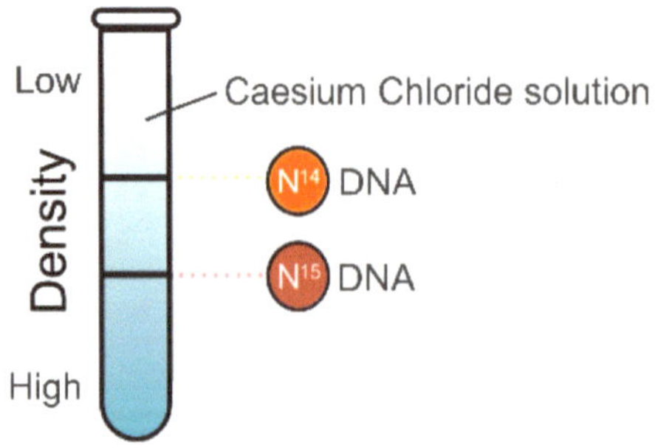

Figure 15.15: Density gradient in Caesium Chloride solution

The experimental techniques and results

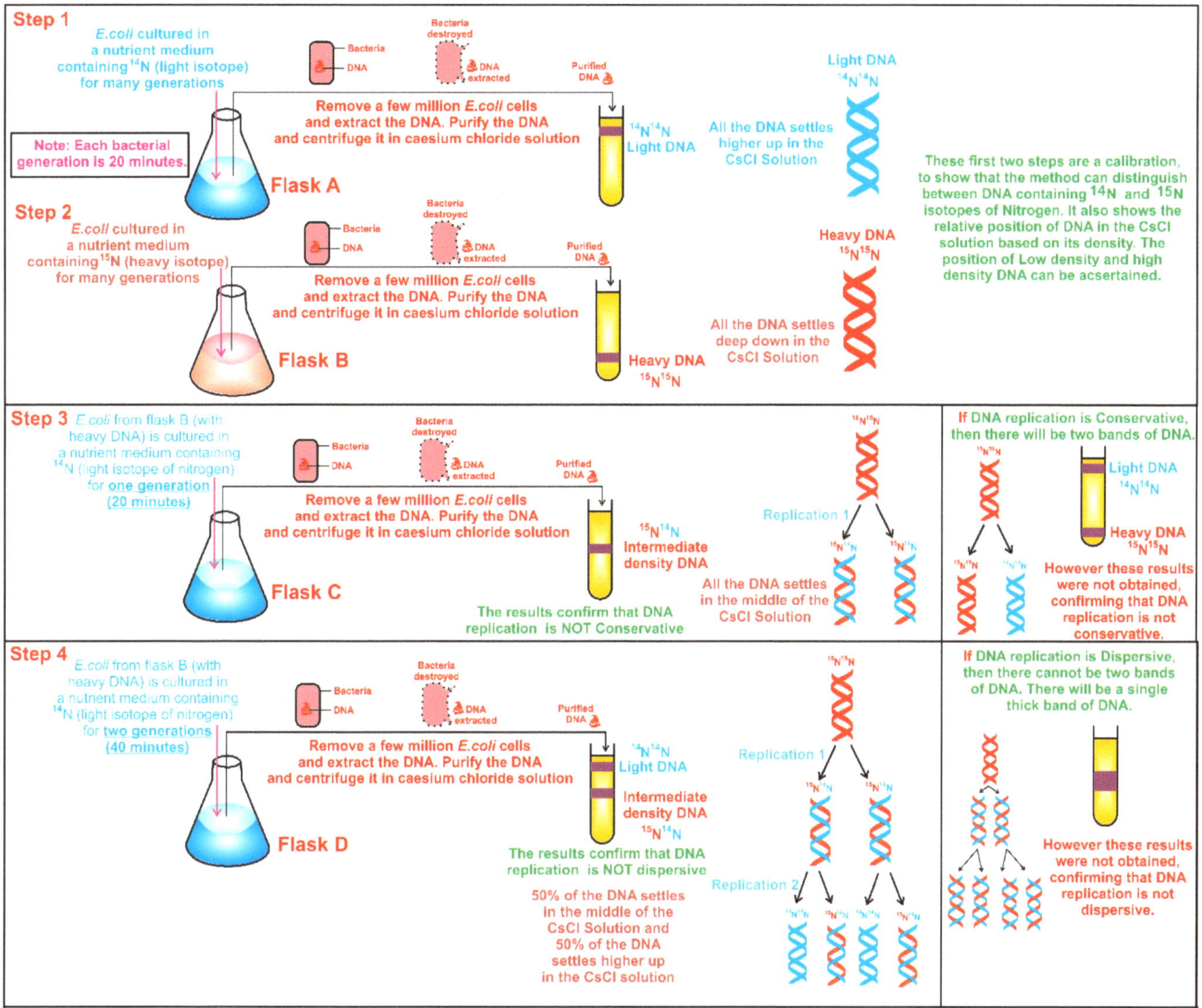

Step 3 indicates that DNA replication is NOT conservative. Step 4 indicates that DNA replication is NOT dispersive.
Hence, the evidence confirms that *DNA replication is semi-conservative*.

Self-assessment (Chapter 15)

Step 1: *Visit www.teachingtube.com*

Step 2: *Sign up for the free course "Edexcel AS and AL Bio assessments"*

Step 3: *Attempt the chapter Quiz and get instant results.*

Step 4: *Download the assessment worksheet and marking scheme too.*

CHAPTER SIXTEEN
THE GENETIC CODE

Edexcel IAL from 2018 Specification 2.11: understand the nature of the genetic code (triplet code, non-overlapping and degenerate)

The Genetic Code – the sequence of bases on Nucleic acids that code for amino acids

The genetic information on the Nucleic acids (DNA and RNA) is in the form of nitrogenous bases. The sequence of bases on DNA determines the sequence of amino acids in a polypeptide chain (primary protein structure). **The sequence of nitrogenous bases which code for amino acids is called as the genetic code**. The fig.18.1shows the codons on **mRNA** that code for specific amino acids.

Second Letter

First Letter	U	C	A	G	Third Letter
U	UUU UUC Phenyl-alanine UUA UUC Leucine	UCU UCC UCA UCG Serine	UAU UAC Tyrosine UAA Stop codon UAG Stop codon	UGU UGC Cysteine UGA Stop codon UGG Tryptophan	U C A G
C	CUU CUC CUA CUG Leucine	CCU CCC CCA CCG Proline	CAU CAC Histidine CAA CAG Glutamine	CGU CGC CGA CGG Arginine	U C A G
A	AUU AUC AUA Isoleucine AUG Methionine (start codon)	ACU ACC ACA ACG Threonine	AAU AAC Asparagine AAA AAG Lysine	AGU AGC Serine AGA AGG Arginine	U C A G
G	GUU GUC GUA GUG Valine	GCU GCC GCA GCG Alanine	GAU GAC Aspartic acid GAA GAG Glumatic acid	GGU GGC GGA GGG Glycine	U C A G

Figure 16.1: The genetic code

Triplet code

If the code is made up of one base it could code for 4 amino acids, as shown in figure 16.2.

Genetic code (Single base)	Amino acid
A	XXX
T	XXX
C	XXX
G	XXX

Figure 16.2: A single base code can code for only 4 amino acids

If the code is made up of two bases it could code for 16 amino acids, as shown in fig.18.3.

Genetic code (Double base)	Amino acid	Genetic code (Double base)	Amino acid	Genetic code (Double base)	Amino acid	Genetic code (Double base)	Amino acid
AA	XXX	TA	XXX	CA	XXX	GA	XXX
AT	XXX	TT	XXX	CT	XXX	GT	XXX
AC	XXX	TC	XXX	CC	XXX	GC	XXX
AG	XXX	TG	XXX	CG	XXX	GG	XXX

Figure 16.3: A double base code can code for only 16 amino acids

Twenty different amino acids are needed to make all the proteins that are needed by living organisms. A single base code or a double base code would not be sufficient enough to code for all the twenty amino acids. A **triplet base code would be able to code for all the 20 amino acids**. So, the bases are read in groups of **3 (Triplet code)**. This gives 64 combinations, more than enough to code for 20 amino acids. The genetic code on the antisense strand of DNA is shown in figure 16.4.

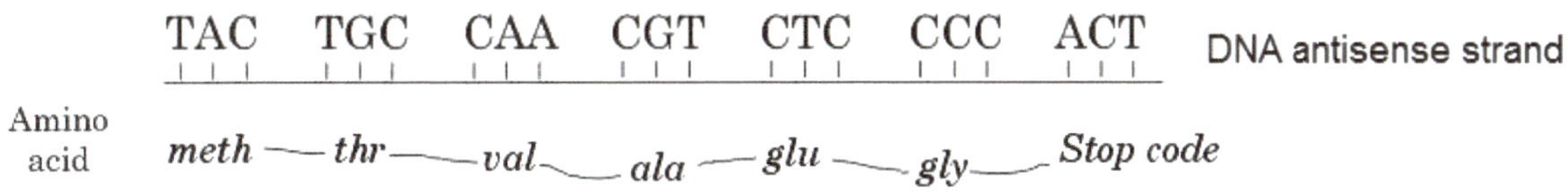

Figure 16.4: Three bases code for one amino acid

Deciphering the code

In the early 1960's Nirenberg devised a method of producing polypeptides by using synthetic mRNA in a cell-free system. The procedure is illustrated in figure 16.5.

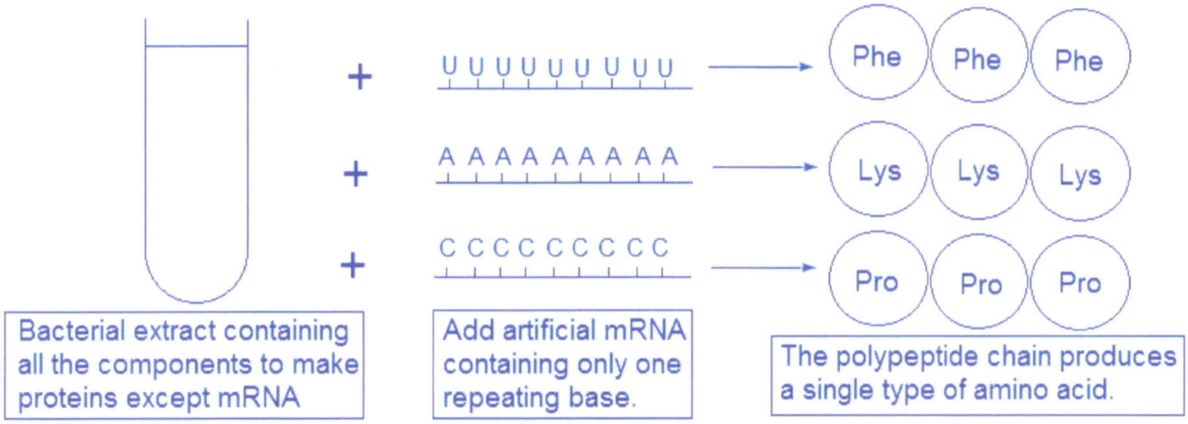

Figure 16.5: Artificial synthesis of polypeptides

The results of this experiment paved the way for determining the genetic code and helped to prove that the genetic code is **a triplet code**, as the number of the **bases on mRNA was three times greater than the number of amino acids in the polypeptide chain**.

Degenerate or redundant code

The genetic code is degenerate - this means that a given amino acid can be coded for by more than one triplet code or codon. eg: GCU, GCC, GCA, GCG all code for the amino acid alanine. Likewise, CAU and CAC code for the same amino acid – Histidine, as shown in figure 16.6.

DNA triplet code	A	G	A	C	A	G	G	T	A	G	T	G	A	T	T
mRNA codon	U	C	U	G	U	C	C	A	U	C	A	C	U	A	A
Amino acid	**Serine**			**Valine**			**Hitstidine**			**Hitstidine**			**STOP Codon**		

Figure 16.6: Degenerate code

The advantage of the degenerate code is that it helps to neutralise mutations. A codon may be changed, but the new codon still codes for the same amino acid. E.g. AUU, AUC and AUA all code for the amino acid Isoleucine. Changing the last letter makes no difference to the polypeptide chain.

Non-overlapping code

The genetic code is described as non-overlapping because **each nucleotide is part of only one triplet codon,** as shown in figure 16.7. There is no overlap with one nucleotide being part of two triplet codes. This means that successive triplets are read in order.

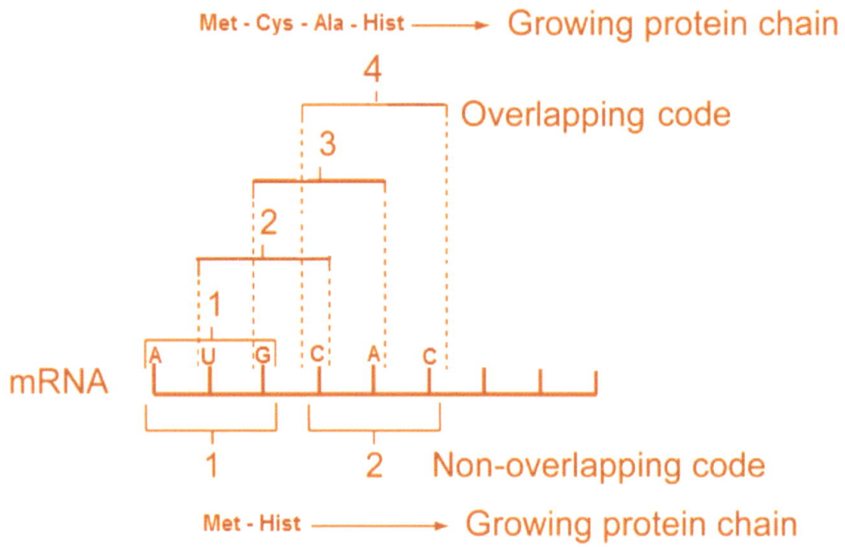

Figure 16.7: Overlapping code

For example, the mRNA sequence AUG CAC could code for two amino acids, methionine (AUG) and histidine(CAC). It DOES NOT code for methionine (AUG), cysteine(UGC), alanine(GCA) and histidine(CAC), as the code is NOT an overlapping code as shown in the upper region of fig 9.7.

An overlapping code is very economical - relatively short lengths of DNA could carry the instructions for many different proteins. However, it would also be very limiting, because the amino acids which could be coded for side by side would be limited. In the example given, only cysteine(UG_) or tryptophan (UG_) out of the 20 available amino acids could ever follow methionine, because only cysteine(UG_) and tryptophan(UG_) has a codon starting with UG-.

Self-assessment (Chapter 16)

Step 1: *Visit www.teachingtube.com*
Step 2: *Sign up for the free course "Edexcel AS and AL Bio assessments"*
Step 3: *Attempt the chapter Quiz and get instant results.*
Step 4: *Download the assessment worksheet and marking scheme too.*

CHAPTER SEVENTEEN
PROTEIN SYNTHESIS

Edexcel IAL from 2018 Specification 2.12: know that a gene is a sequence of bases on a DNA molecule that codes for a sequence of amino acids in a polypeptide chain

Gene

A gene is **a sequence of bases on a DNA molecule coding for a sequence of amino acids in a polypeptide chain**.

Genes express themselves by specifying the sequence of amino acids in polypeptide chains during **protein synthesis,** as shown in figure 17.1.

Figure 17.1: Protein synthesis

The genes control metabolic activity by producing specific **enzymes** needed for metabolism. It may also produce specific **functional proteins** like Insulin, Haemoglobin, Myoglobin, Rhodopsin and Glucagon. Genes also specify the amino acid sequence of **structural proteins** like keratin (in nails), collagen (in bones), actin and myosin (in muscles), carrier and channel proteins (in cell membranes).

- DNA consists of several genes along its length. The function of the gene is to provide information for building proteins.
- Figure 17.2 shows that gene A codes for polypeptide chain A, gene B codes for polypeptide chain B, and gene C codes for polypeptide chain C.
- The sequence of bases in each gene determines the sequence of amino acids in the polypeptide chain.
- The regions of DNA which do not code for polypeptides is called as 'junk'. These regions are used for regulating the 'switching on' and 'switching off' of genes.

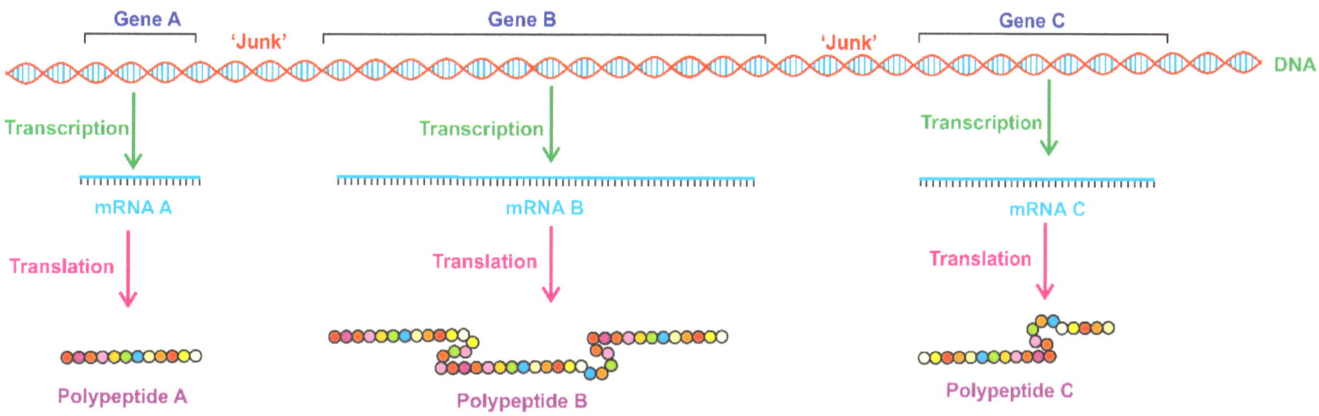

Figure 17.2: Overview of protein synthesis

Edexcel IAL from 2018 Specification 2.13: (i) understand the process of protein synthesis (transcription and translation), including the role of RNA polymerase, translation, messenger RNA, transfer RNA, ribosomes and the role of start and stop codons
(ii) understand the roles of the DNA template (antisense) strand in transcription, codons on messenger RNA and anticodons on transfer RNA

Protein synthesis

Protein synthesis is the formation of a polypeptide chain by using the information on the DNA. The information from the gene (DNA) is copied on to mRNA in the nucleus by a process called **transcription**. The mRNA is then used on the ribosomes to form a polypeptide chain by a process called **translation**.

Transcription - making mRNA from DNA

1. A length of DNA, called as a gene or cistron, is copied onto a single stranded mRNA molecule during transcription.

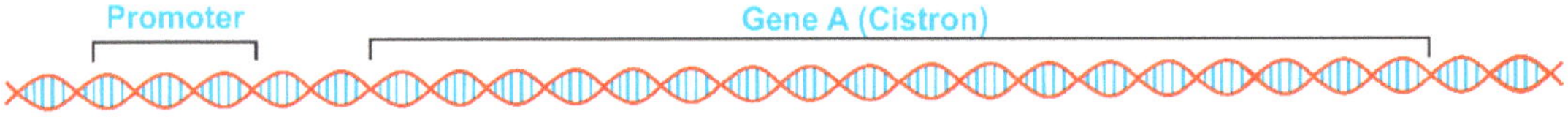

2. The enzyme RNA polymerase and transcription factors attach to the promoter region of DNA. This activates the gene and switches it on. The RNA polymerase can now start to unzip the gene.

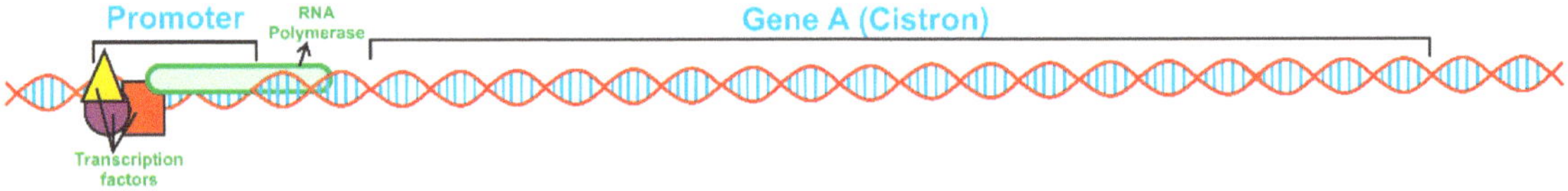

3. RNA polymerase moves along the gene and begins to unwind the **gene** (cistron) by breaking the hydrogen bonds between the nitrogenous base pairs. Free ribonucleotides in the nucleus pair up with complementary bases on the **antisense strand** of DNA by forming hydrogen bonds. **A** pairs with **U**; **G** pairs with **C; C** pairs with **G; T** pairs with **A.**

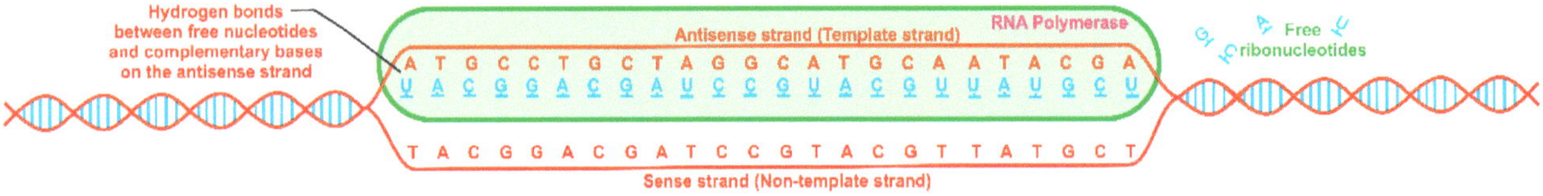

4. The free nucleotides then get linked to each other by the formation of phospho-diester bonds, formed by condensation reactions. It is catalysed by RNA polymerase.

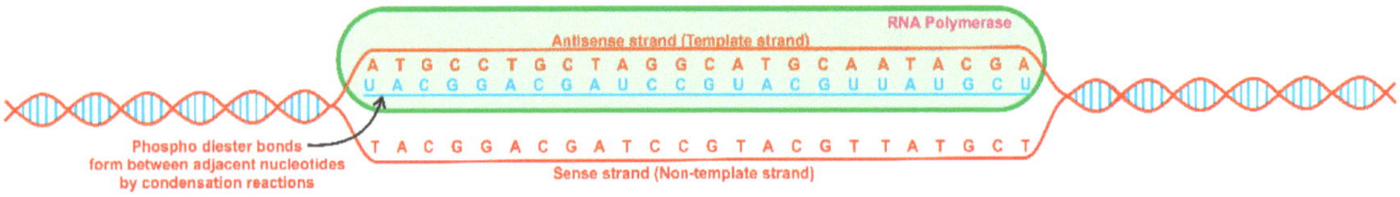

5. The newly formed strand of mRNA then peels away from the coding strand and the DNA rewinds. The process continues until a stop codon is reached.

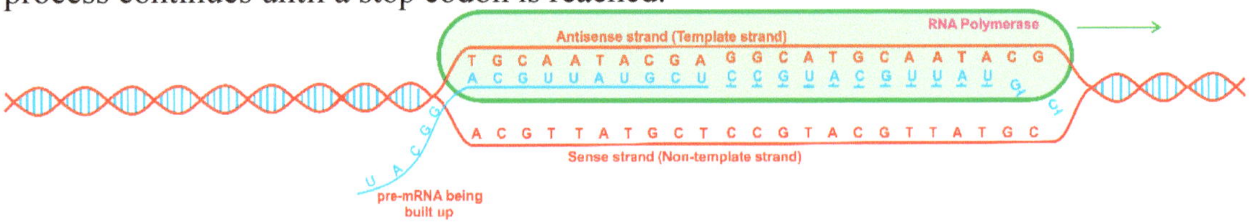

6. A primary transcript of mRNA or pre-mRNA is formed. It contains introns and exons.

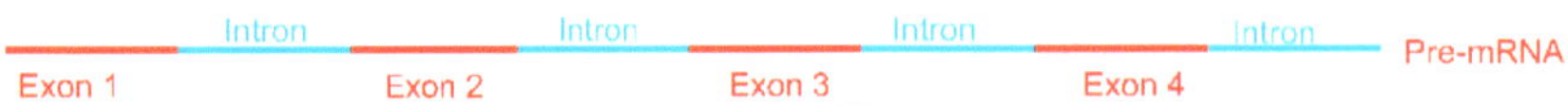

7. Before leaving the nucleus, introns of mRNA are cut off by a process called splicing. The remaining nucleotides (exons) rejoin to form mRNA.

8. Poly adenine tail and guanine cap are attached to the mRNA before it diffuses out of the nucleus through the nuclear pores.

Translation – decoding genetic information on mRNA to form a polypeptide

During translation the sequence of bases on mRNA is used to form a polypeptide chain with a specific sequence of amino acids. It occurs on the ribosomes, as describe below.

1. The smaller sub-unit of ribosome attaches to the 5' end of mRNA at an initiation codon (AUG).

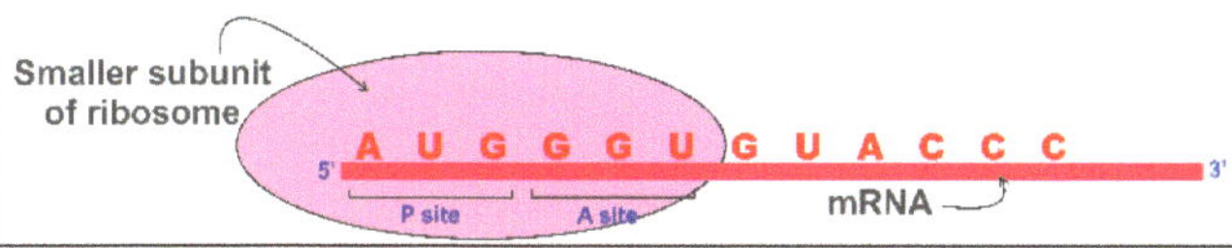

2. The anticodon (UAC) of a transfer RNA carrying a methionine binds to the AUG codon on the messenger RNA, by complementary base pairing.

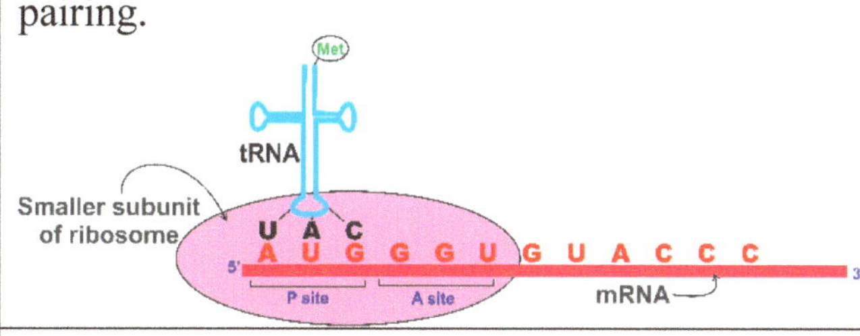

3. The larger subunit of the ribosome then attaches to the system.

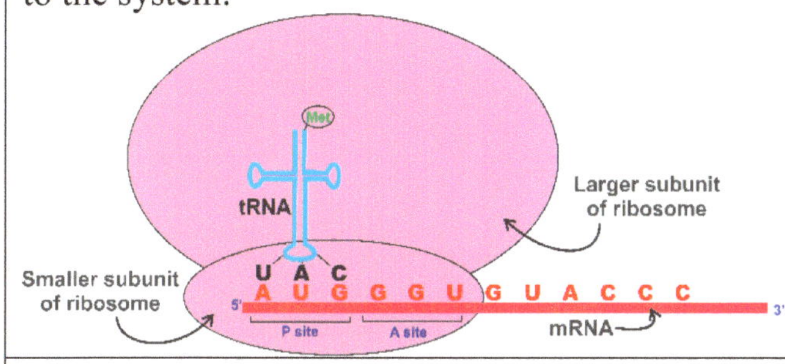

4. The next amino acid-tRNA attaches to the adjacent mRNA codon (Gly in this case).

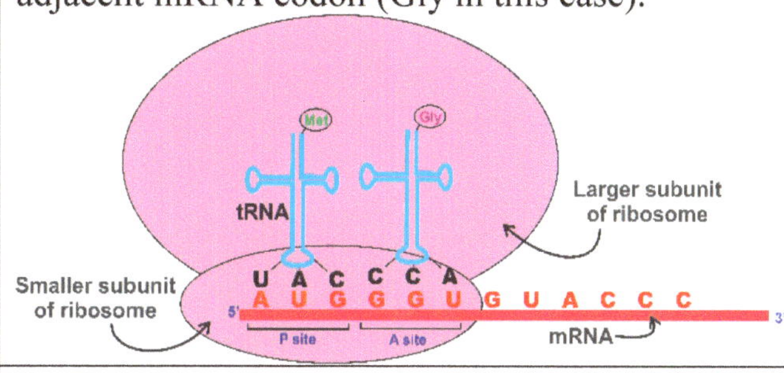

5. The bond between the tRNA and the amino acid breaks and a <u>peptide bond</u> is formed between the two amino acids at the P site and the A site. The enzyme peptidyl-transferase is involved in formation of the peptide bond.

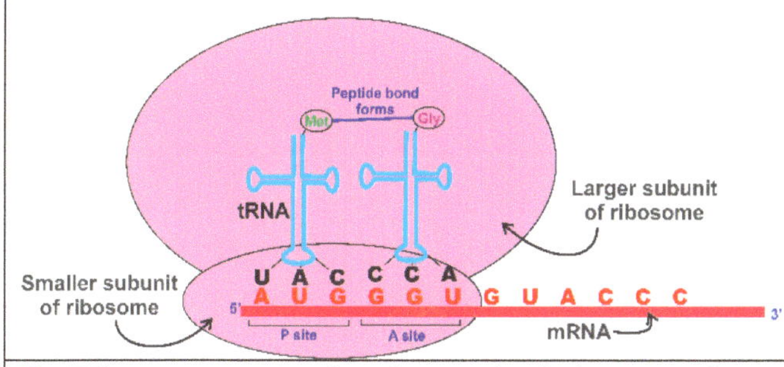

6. The free tRNA molecule leaves to collect another amino acid. The polypeptide chain elongates one amino acid at a time, and peels away from the ribosome, folding up into a protein as it goes. This continues until a stop codon is reached.

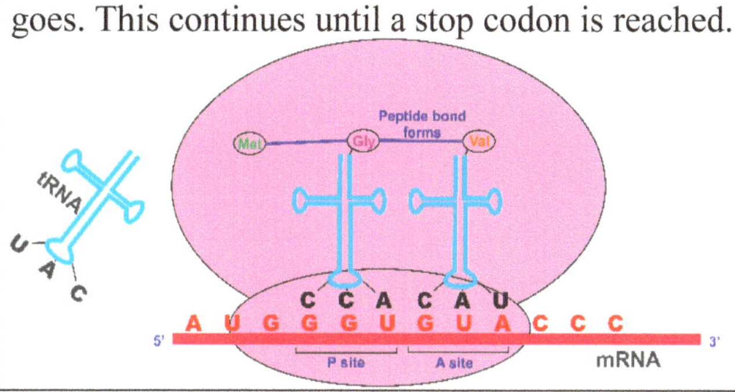

7. When a stop codon is reached then a release factor binds to the stop codon.

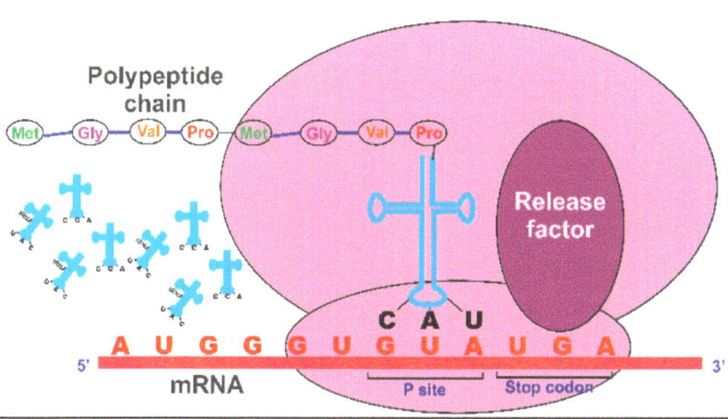

8. The release factor separates the two ribosomal sub-units and releases the polypeptide chain and mRNA.

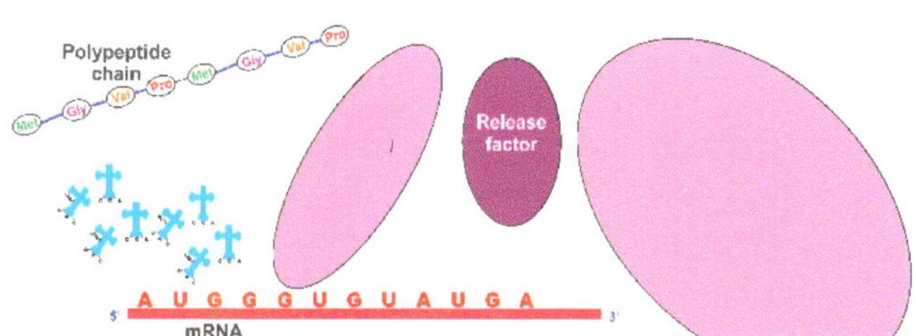

A single piece of mRNA can be translated by many ribosomes simultaneously. A group of ribosomes all attached to one piece of mRNA is called a polysome, as shown in figure 17.3.

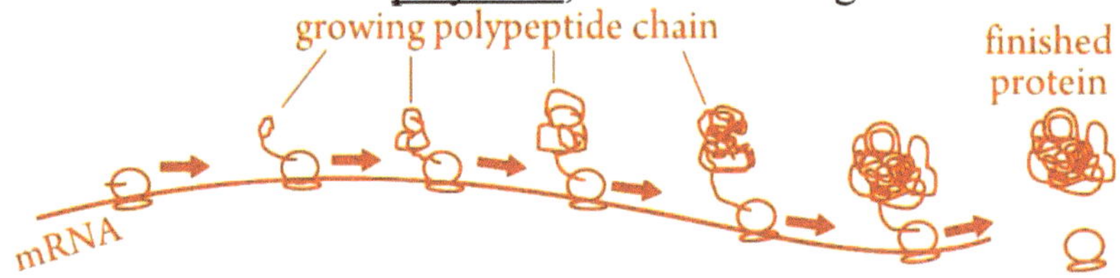

Figure 17.3: A polysome

*Note: The m*ain function of the ribosome is to hold the mRNA molecule so that anticodons of tRNA can pair up with complementary codons of mRNA. This brings amino acids to lie adjacent to each other so that peptide bonds can be formed.

Structure of transfer RNA (tRNA)

tRNA is only about 80 nucleotides long and it folds up by complementary base pairing to form a clover-leaf structure, as shown in figure 17.4. At one end of the molecule there is an **amino acid binding site**. On the middle loop there is a triplet nucleotide sequence called the **anticodon**. There are 61 different tRNA molecules, each with a different anticodon sequence complementary to the 61 different codons on mRNA (there is no pairing of tRNA with the stop codons).

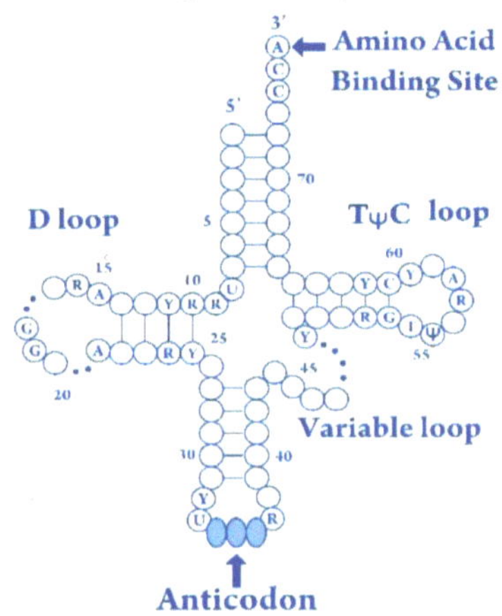

Figure 17.4: Structure of tRNA

The tRNA binds with a specific amino acid in the cytoplasm to form an amino-acyl tRNA complex, using the enzyme amino-acyl tRNA synthetase. ATP is also used in this process. The activated amino acid is then transferred to the ribosome.

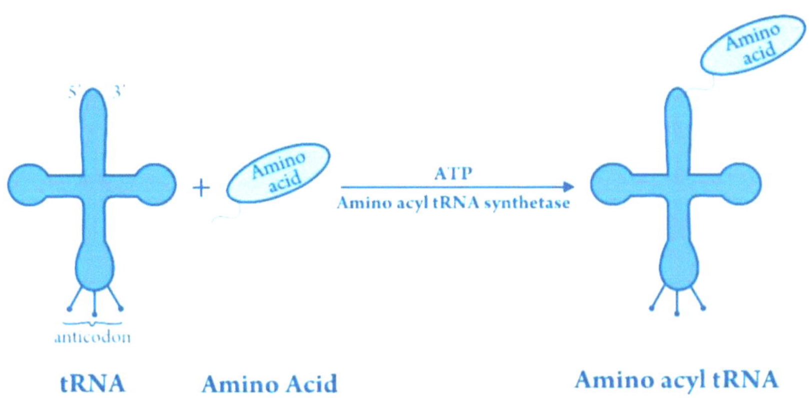

Figure 17.5: Role of tRNA

The amino acid with which tRNA binds is determined by the anticodon of the tRNA. This enables the anticodon of tRNA to pair up with complementary codons of mRNA and build up a specific sequence of amino acids in the polypeptide chain.

Role of anticodon

The anticodon is a sequence of three bases that will pair up with complementary codons of mRNA.
The anticodon determines which amino acid can bind with the specific tRNA molecule. For example, the tRNA carrying the anticodon UAC can bind with the amino acid methionine only. The tRNA matches the specific amino acid to its codon, by complementary base pairing between the codon and anticodon as shown in figure 17.6.

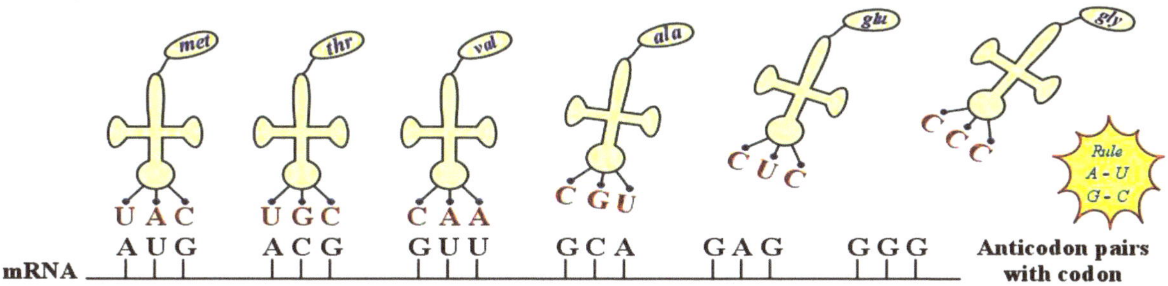

Figure 17.6: Role of anticodon on tRNA

Note: anti-codons can pair with complementary codons only within the ribosomes.

Role of start codon - finding the start point of the messenger RNA and initiating translation

Ribosomes come in two parts - a smaller subunit and a larger subunit. The smaller subunit is involved in finding the start point. It attaches to the 5' end of the messenger RNA and binds to the AUG codon, as shown in figure 17.7.

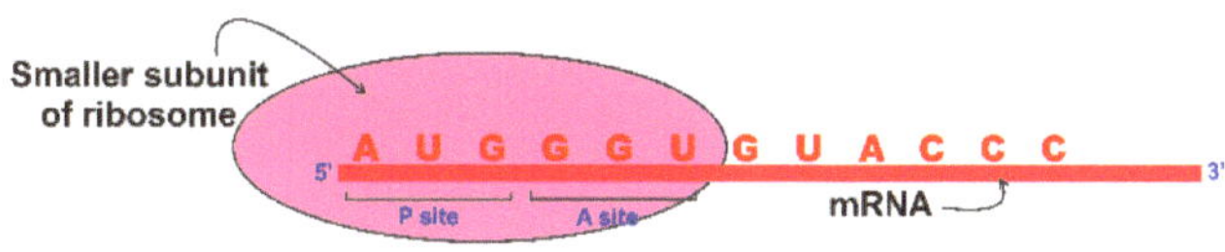

Figure 17.7: Smaller subunit of ribosome attaches to the start codon

Now two things happen.

1. The anticodon (UAC) of a transfer RNA carrying a methionine binds to the AUG codon on the messenger RNA, by complementary base pairing as shown in figure 17.8.

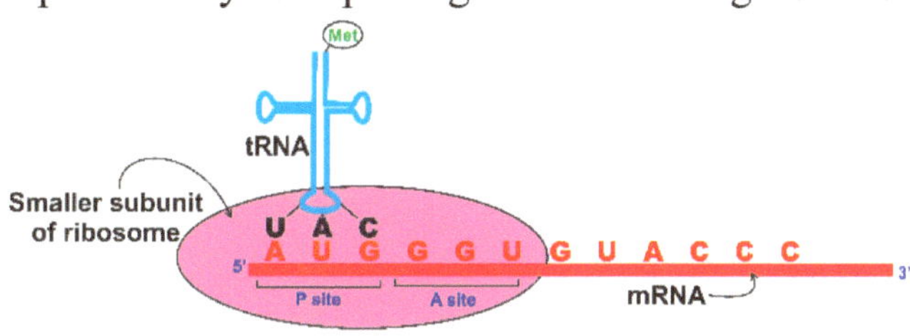

Figure 17.8: Pairing of UAC anticodon with the start codon

2. The larger subunit of the ribosome then attaches to the system as shown in figure 17.9.

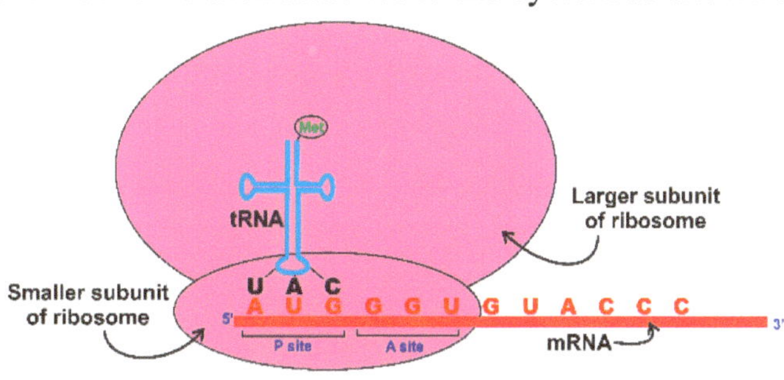

Figure 17.9: Attachment of the larger ribosomal subunit

The process of translation can then occur as explained earlier, by elongation of the chain.

Role of the stop codon

The stop codon signals the **termination of translation**. The polypeptide chain is no longer elongated. The stop codon allows a release factor to bind to the mRNA, as shown in figure 17.10. This prevents another amino acid-tRNA from binding to the chain and will cause the two ribosomal sub-units to separate. The polypeptide chain is released from the ribosome, along with release factor, when the ribosomal subunits separate, as shown in figure 17.11.

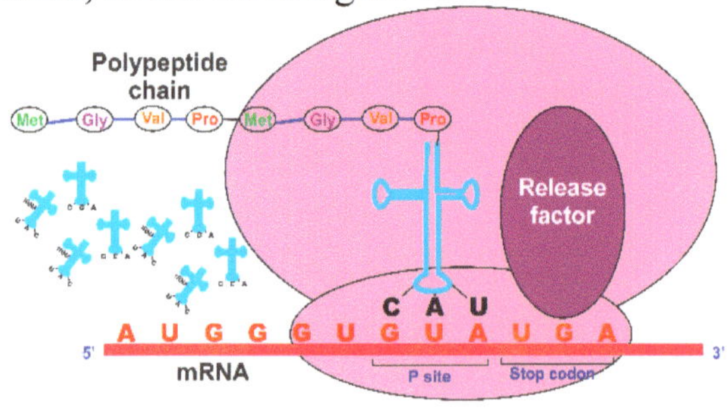

Figure 17.10: Attachment of the release factor to the stop codon

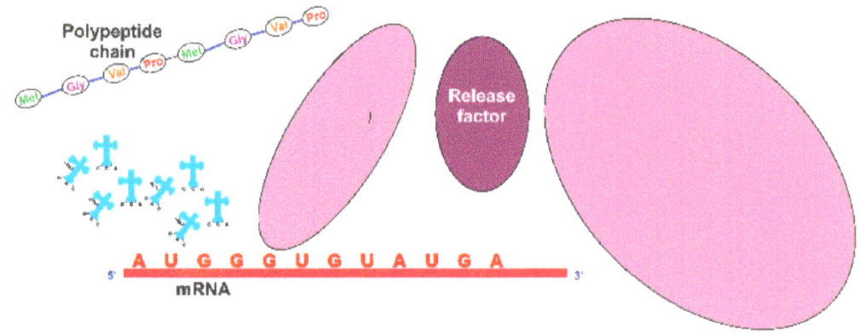

Figure 17.11: Ribosomal subunits separate

Role of the antisense strand of DNA

The antisense strand serves as a template for the formation of mRNA. The mRNA is complementary to the (template) antisense strand. The information from the antisense strand is copied on to mRNA during transcription, as shown in figure 17.12.

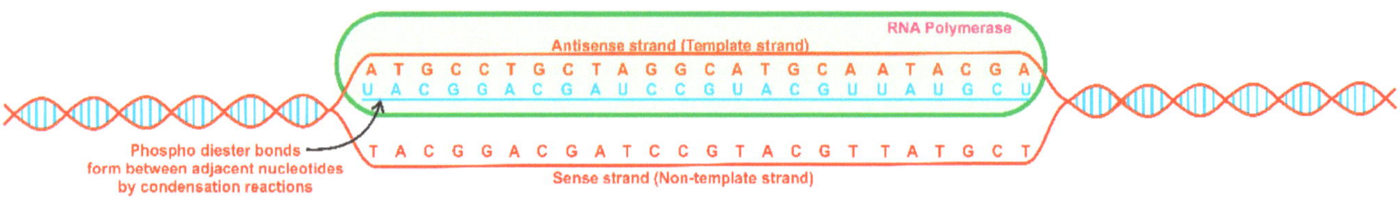

Figure 17.12: Antisense strand is a template for the formation of mRNA

Role of the sense strand of DNA

The sense strand (non-template) of DNA is similar to the mRNA (except that T is replaced with U), which codes for the amino acids in the polypeptide chain. It contains the sequence of bases to which RNA polymerase binds and begins transcription. The sense strand is also referred to as the *coding* strand (because it is a DNA copy of the RNA sequence)

Role of mRNA *(messenger* RNA) codons

mRNA is a complementary copy of the antisense strand of DNA. Its function is to **copy and carry** genetic information from the DNA inside the nucleus to the ribosomes in the cytoplasm for proteins synthesis (translation). The genetic code can also be read from mRNA strand. A sequence of three nitrogenous bases on mRNA which codes for a specific amino acid is called a **Codon**.

Figure 17.13 shows the relationship between the **triplet bases** on DNA, the **codons** on mRNA and the amino acids that are coded for by the respective bases.

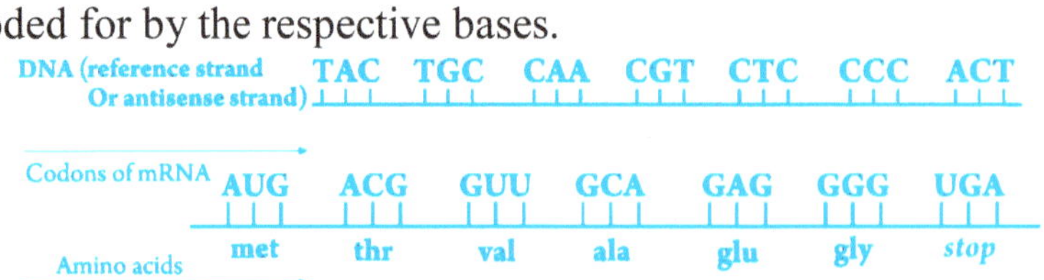

Figure 17.13: The sequence of bases on mRNA determines the sequence of amino acids

Self-assessment (Chapter 17)

***Step 1:** Visit www.teachingtube.com*

***Step 2:** Sign up for the free course "Edexcel AS and AL Bio assessments"*

***Step 3:** Attempt the chapter Quiz and get instant results.*

***Step 4:** Download the assessment worksheet and marking scheme too.*

CHAPTER EIGHTEEN

MUTATIONS

Edexcel IAL from 2018 Specification 2.14: **(i) understand how errors in DNA replication can give rise to mutations (substitution, insertion and deletion of bases)**
(ii) know that some mutations will give rise to cancer or genetic disorders, but that many mutations will have no observable effect

Mutation

A mutation is a **heritable change in the DNA of a cell**. Gene or **point mutations** result from a **change in the base sequence within a gene, usually during DNA replication**.

If mutations occur in **gamete cells, zygote, precursor cells of gametes**, then they can be passed on to successive generations. This is called as **germ line mutations**. Gene mutations may result in genetic disorders like, cystic fibrosis, Albinism and thalasaemia.

Mutations that occur in body cells cannot be passed onto successive generations. These mutations can only be passed onto daughter cells formed by mitosis from the mutant cell. These are called as **somatic mutations.**

Mutations and cancer

Most of the cells with mutations will die out and have no adverse effect on the health of the individual. However, mutations in tumor suppressor genes or oncogenes (genes that regulate mitosis) can lead to cancer.

Mutations occur **randomly and spontaneously** in nature, during DNA replication. The frequency with which one allele mutates to another is known as the **mutation rate**.

The flowchart below illustrates the effects of gene mutations.

Change in bases of DNA → Change in bases of mRNA → Change in mRNA codon → Change in amino acid of a protein → Change in protein structure → Change in protein function → Change in cell function

A few types of gene mutations are described below

Gene or point mutations result from a change in the base sequence within a gene. Remember that a gene is a part of the DNA that codes for a sequence of amino acids in a polypeptide chain.

Point mutations can occur by:

Substitution

One base in the DNA sequence is replaced by another base.

The effect of a substitution mutation is illustrated in figure 18.1.

	val	*his*	*leu*	*thr*	*pro*	*glu*	*glu*	*lys*	*Normal sequence of the first 8 amino acids in the beta globin chain*
Normal DNA	*CAA*	*GTA*	*AAT*	*TGA*	*GGA*	*CTC*	*CTC*	*AAA*	*DNA segment coding for normal haemoglobin*
Mutant DNA	*CAA*	*GTA*	*AAT*	*TGA*	*GGA*	*CAC*	*CTC*	*AAA*	*DNA segment coding for sickle cell haemoglobin*
	val	*his*	*leu*	*thr*	*pro*	*val*	*glu*	*lys*	*Defective beta globin chain in sickle cell haemoglobin*

Figure 18.1: A substitution mutation

Deletion

This is a point mutation brought about by removal of one base from a gene.

Insertion

This mutation involves the insertion or addition of a base into a gene.

Substitution results in one or two amino acids being altered in the polypeptide chain. However, deletion and insertion result in a complete change of the amino acid sequence in the polypeptide chain. This is because, the addition or removal of a single base, causes a shift in the whole sequence of bases and all the codons are changed after that point. So, insertion and deletion are far more damaging than substitution. A sequence of 3 bases is called is a frame. Addition or removal of a single base from the sequence will change all the frames beyond the point of insertion or deletion, as shown by the simple example in figure 18.2 and 18.3.

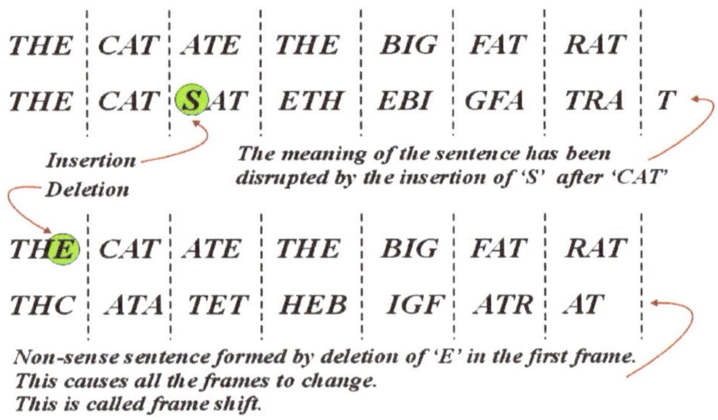

Figure 18.2: Insertion or deletion of one letter changes all words

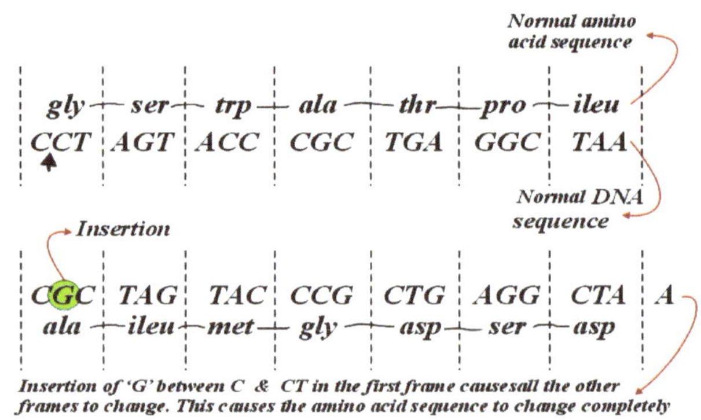

Figure 18.3: Insertion or deletion of a base changes all the amino acids

Neutral Mutations: The sequence of bases changes but the amino acid sequence remains unchanged. So, the is no effect on the phenotype.

Neutral mutations may occur in two ways:

a.) A codon is changed, but the **new codon still codes for the same amino acid** e.g. AUU, AUC and AUA all code for the amino acid isoleucine. Changing the last letter makes **no difference in the polypeptide chain**.

b.) A codon is changed resulting in a new amino acid in the protein. However, the different amino acid **does not change the shape of the protein**, and the activity of the protein is unaffected.

Self-assessment (Chapter 18)

Step 1: Visit www.teachingtube.com
Step 2: Sign up for the free course "Edexcel AS and AL Bio assessments"
Step 3: Attempt the chapter Quiz and get instant results.
Step 4: Download the assessment worksheet and marking scheme too.

Chapter Nineteen
Genetics

Edexcel IAL from 2018 Specification 2.15: (i) understand what is meant by the terms *gene, allele, genotype, phenotype, recessive, dominant, codominance, homozygote* and *heterozygote*
(ii) understand patterns of inheritance, including the interpretation of genetic pedigree diagrams, in the context of monohybrid inheritance.
(iii) understand sex linkage on the X chromosome, including red-green colour blindness in humans

Monohybrid inheritance

The monohybrid cross involves the inheritance of a single pair of genes. The gene usually has two alleles. One allele is dominant and the other is recessive.

Consider an example of a monohybrid cross involving the inheritance of shape of pea seeds, illustrated in figure 19.1. This gene has two alleles - Round (**R**) and Wrinkled (**r**).

It is conventional to represent the dominant allele by a capital letter and the corresponding recessive allele by the simple form of the same letter.

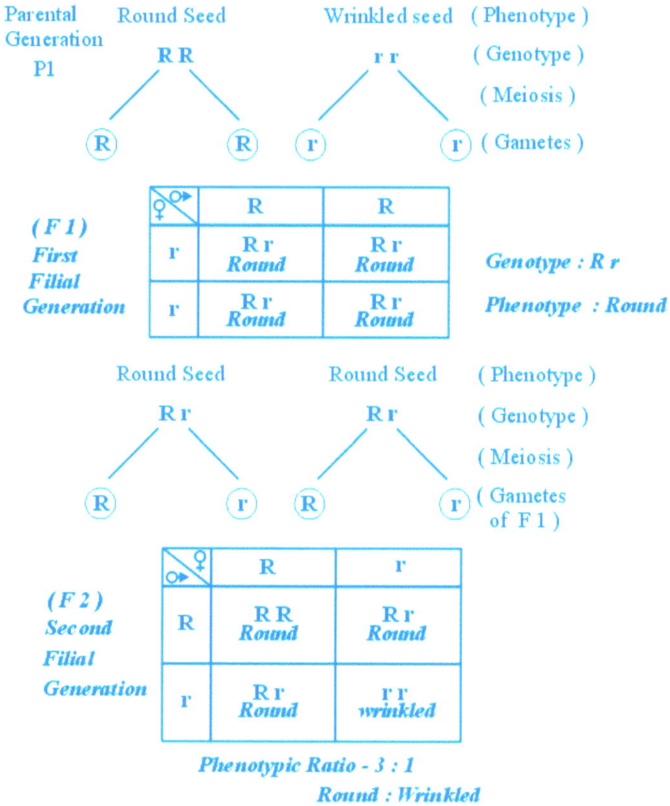

Figure 19.1: Monohybrid cross

Gene

A gene is a sequence of bases on DNA that codes for a single polypeptide chain. The gene locus is the location of a gene on the chromosome.

Allele

A gene can exist in two or more different forms. These **alternate forms of a gene found at a gene locus are referred to as alleles.** The alleles differ from each other in the sequence of bases.

It is conventional to represent the dominant allele by a capital letter and the corresponding recessive allele by the simple form of the same letter. For example, the gene for **seed shape in pea plants** can exist in two forms: **R** (allele for round seed) or **r** (allele for wrinkled seed). Thus, the alleles of the gene for seed shape in pea plants are **R** and **r**.

Genotype
The combination of alleles found at the same gene locus of homologous chromosomes. For example, the genotypes RR, Rr and rr in pea plant inheritance of seed shape.

Homozygote
A genotype consisting of similar alleles. For example, in the inheritance of seed shape in pea plants the genotypes TT and tt are homozygous.

Heterozygote
A genotype consisting of two different alleles. For example, in the inheritance of seed shape in pea plants the genotypes Tt is heterozygous.

Phenotype
The observable features of an organism that are produced by the interaction of the genotype and the environment. For example, round seeds or wrinkled seeds are the observable features of pea plants.

Recessive allele
The allele which expresses itself in the phenotype, only in homozygous conditions, but, not in heterozygous conditions. For example, the allele r produces wrinkled seeds only in homozygous (rr) condition. In heterozygous condition (Rr), the recessive allele (r) is present, but is not expressed in the phenotype.

Dominant allele
The allele which expresses itself in the phenotype, both, in homozygous and in heterozygous genotypes. For example, the allele R expresses itself when found as RR to give rise to tall pea plants, and when found as Rr it produces tall pea plants.

Codominance: When both alleles are expressed in the phenotype during heterozygous condition.

Patterns of inheritance
Humans have 23 pairs of chromosomes, as shown in figure 19.2.

- 22 pairs of **autosomes** (which do not play a role in sex determination), and
- 1 pair of **sex chromosomes** (which play a role in sex determination).
- **Autosomal disorders** are caused by genes on the autosomes and can affect both males and females indiscriminately (equally).
- **Sex–linked disorders** are caused by genes located on the sex chromosomes.

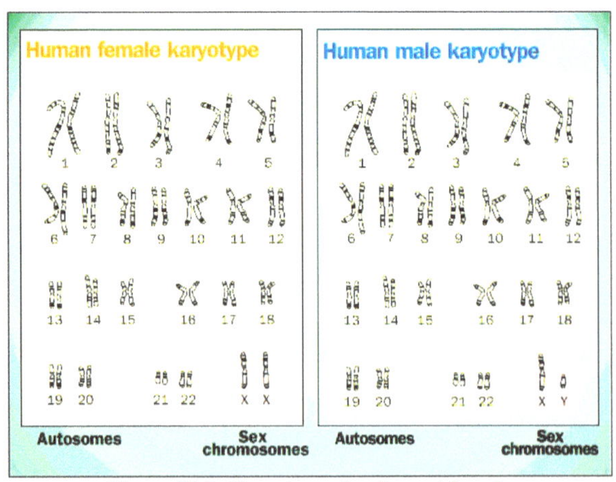

Figure 19.2: Karyotype of males and females

All people have the same genes, which code for the proteins that make up our body. Because we have 2 copies of each chromosome we have 2 copies of each gene. However, it is possible to have different versions of a gene. The versions are called **alleles**. Because alleles are different versions of the **same gene** they are found in the **same position** on each of the pairs of chromosomes. The position that a gene occupies on a chromosome is called a **Locus** (latin = place).

Genetic disorders in humans can be caused by dominant or by recessive alleles. The inheritance and expression of the disorder can also be influenced by the location of the gene, i.e. whether it is found on the autosomes or on the sex chromosomes.

Some genetic conditions are caused by mutations in a single gene. These conditions are usually inherited in one of several patterns, depending on the gene involved.

Autosomal inheritance – genes found on autosomes

Autosomal dominant disorders

- The disorder is caused by a dominant allele (A) of the gene.
- The recessive allele (a) produces the normal protein.
- A homozygous dominant (AA) or heterozygous (Aa) individual will suffer from the genetic disorder.
- The affected person may inherit the condition from an affected parent, or the condition may result from a new mutation in the gene and occur in people with no history of the disorder in their family. In Huntington disease there are no symptoms until the late 30s. The consequence of this is that most people would have already had children and unwittingly passed on the mutant allele to their children.
- Examples: Huntington disease, Marfan syndrome

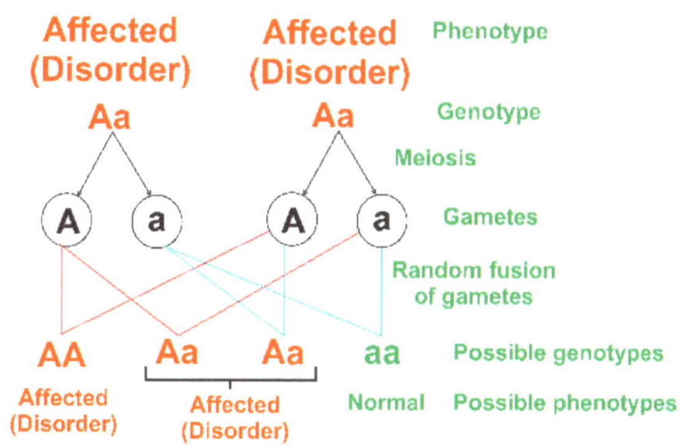

Figure 19.3: Inheritance of autosomal dominant disorders

Autosomal recessive disorders

- The disorder is caused by a recessive allele (a) of the gene.
- The dominant allele (A) produces the normal protein.
- A homozygous dominant (AA) or heterozygous (Aa) individual will be normal.
- Only homozygous individuals (aa) will suffer from the genetic disorder.
- The parents of an individual with an autosomal recessive condition each carry one copy of the mutated gene, but they typically do not show signs and symptoms of the condition.
- Example: Cystic fibrosis, Albinism, Thalassemia, Sickle cell anaemia

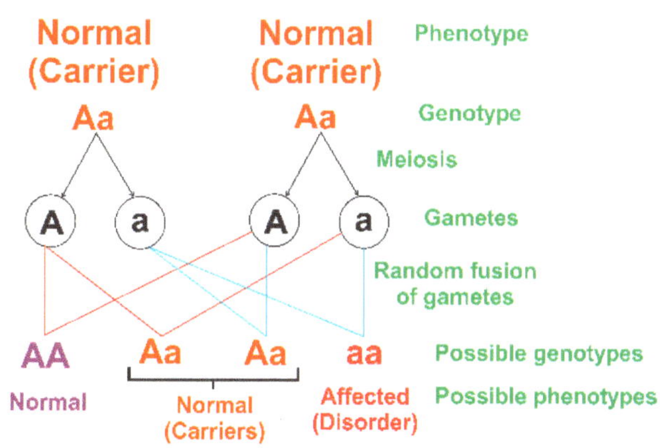

Figure 19.4: Inheritance of autosomal recessive disorders

Sex-linked inheritance – genes found on sex chromosomes

When assigning alleles for a sex-linked trait, the convention is to write the allele as a superscript to the sex chromosome (X). Example: X^A, X^a, Y^A, Y^a, Y^o.

X-linked dominant disorders

- X-linked dominant disorders are caused by mutations in genes on the X chromosome.
- In females (who have two X chromosomes), a mutation in one of the two copies of the gene in each cell is sufficient to cause the disorder (X^AX^A) or (X^AX^a).
- In males (who have only one X chromosome), a mutation in the only copy of the gene in each cell causes the disorder (X^AY^o).
- In most cases, males experience more severe symptoms of the disorder than females.
- A characteristic of X-linked inheritance is that fathers cannot pass X-linked traits to their sons (no male-to-male transmission).
- Example: Fragile X syndrome, Rett syndrome.

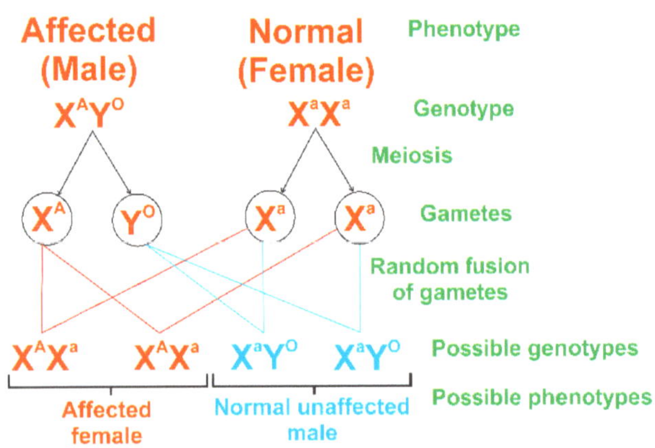

Figure 19.5: Inheritance of X-linked dominant disorders

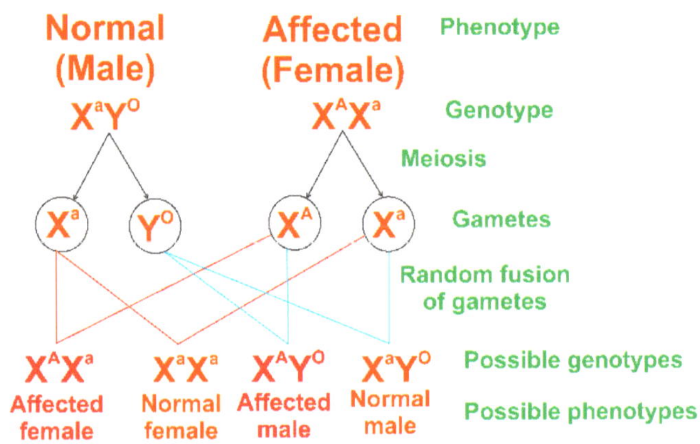

Figure 19.6: Inheritance of X-linked dominant disorders

X-linked recessive disorders

- X-linked recessive disorders are also caused by mutations in genes on the X chromosome.
- In males (who have only one X chromosome), one altered copy of the gene (X^a) in each cell is sufficient to cause the condition.
- In females (who have two X chromosomes), a mutation would have to occur in both copies of the gene (X^aX^a) to cause the disorder.
- Because it is unlikely that females will have two altered copies of this gene, males are affected by X-linked recessive disorders much more frequently than females.
- A characteristic of X-linked inheritance is that fathers cannot pass X-linked traits to their sons (no male-to-male transmission).
- Example: Haemophilia, Red-green colour blindness, Fabry disease

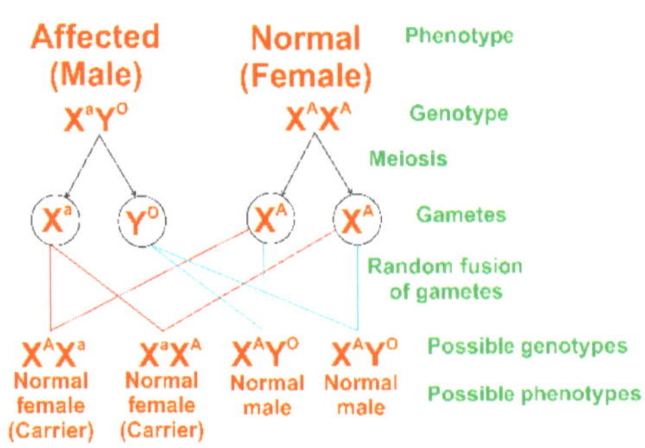

Figure 19.7: Inheritance of X-linked recessive disorders

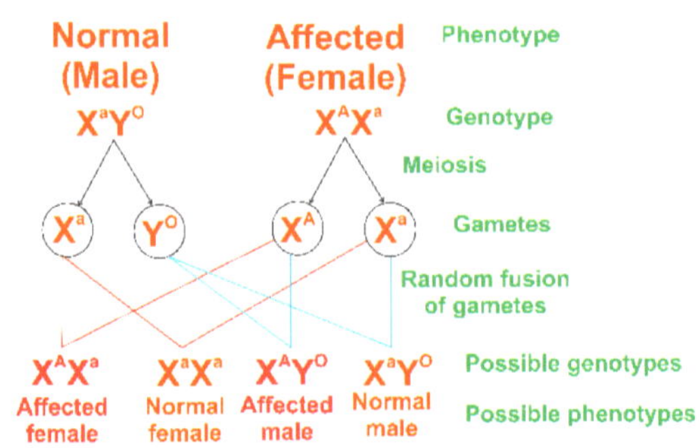

Figure 19.8: Inheritance of X-linked recessive disorders

Red-Green Colour Blindness

Red-green colour blindness is a genetic disorder whereby an individual fails to discriminate between red and green hues. This condition is caused by a mutation to the red or green retinal photoreceptors, which are located on the X chromosome. It is caused by a recessive allele (X-linked recessive disorder). Red-green colour blindness can be diagnosed using the Ishihara colour test.

The Ishihara Colour Test

What Red-Green Colour Blind People See | What People with Regular Vision See

Figure 19.9: The Ishihara colour test chart for red-green colour blindness

Codominant disorders

- In codominant inheritance, two different versions (alleles) of a gene are expressed, and each version makes a slightly different protein.
- Both alleles influence the genetic trait or determine the characteristics of the genetic condition.
- For example, in sickle cell anemia the red blood cells become thin and stretched out.
- Hb^N allele makes normal haemoglobin, and
- Hb^S allele makes defective haemoglobin that results in sickle shaped red blood cells.
- If a person has a single copy of the sickle cell allele (Hb^S), then half their red blood cells become abnormally shaped. If this happens, codominance occurs because both normal and sickled shapes are mixed and seen in the blood.
- The genotypes and phenotypes of sickle cell inheritance are shown below.

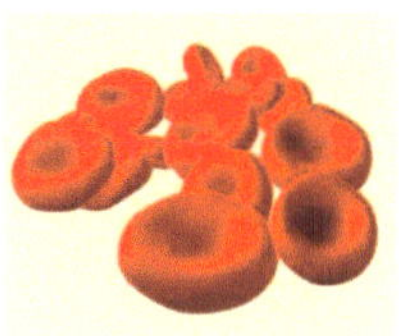 **Hb^N Hb^N** **Homozygous dominant** • Normal phenotype. • All healthy RBCs.	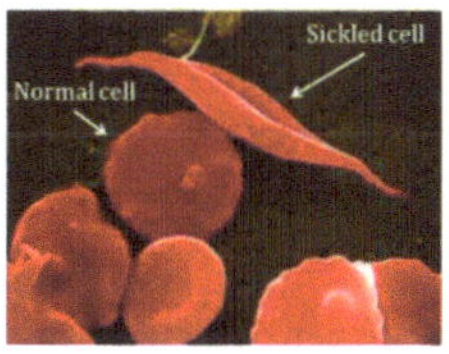 **Hb^N Hb^S** **Codominant (both alleles expressed)** • Normal phenotype but a carrier of the sickle cell allele. • The blood has both type of RBCs.	 **Hb^S Hb^S** **Homozygous recessive** • All the RBCs are sickle shaped. • The person suffers from sickle-celled anaemia. • This is also an example of an **autosomal recessive disorder** too.

Symbols used in family trees (pedigree charts) and genetic diagrams

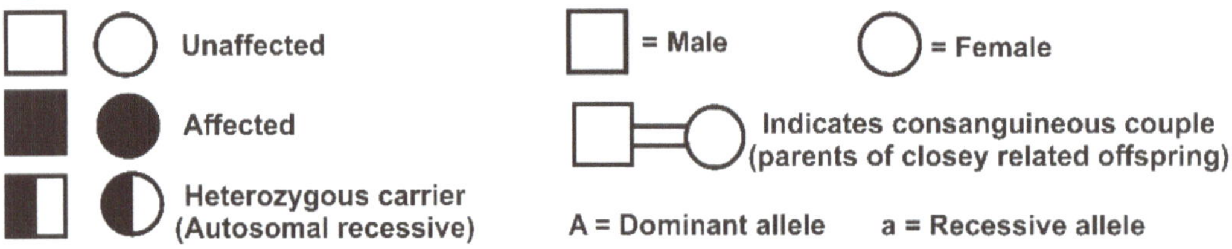

Rules for pedigree charts

- Individuals in the same generation are placed in the same row. For example, Vazna, Amir, Vulau and Shifa all belong to the same generation. Elaf, Sophia, Fizal and Faraha belong to the next generation.
- The older offspring is placed to the right of the younger offspring in each family. For example, Vazna is younger than Amir and Vulau is younger than Shifa.

Interpretation of pedigree charts (Mendelian inheritance)

Autosomal recessive disorders

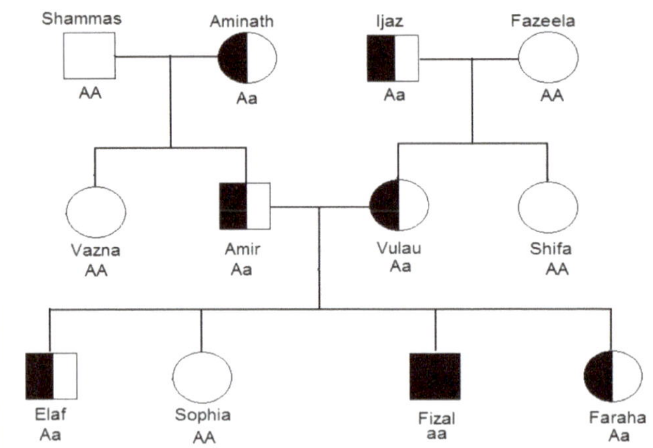

- Vazna (**AA**) has inherited one **A** allele from each parent. So, Vazna will be normal.
- Amir (**Aa**) has inherited one dominant (**A**) allele from Shammas and one recessive (**a**) allele from Aminath. So, Amir is a genetic carrier and can pass on the recessive (disease causing allele) allele to his offspring (Fizal and Elaf). But, Amir is phenotypically normal.
- Fizal (**aa**) has inherited one recessive (**a**) allele from each parent (Amir and Vulau). So, Fizal suffers from a genetic disorder caused by recessive alleles.

Autosomal dominant disorders

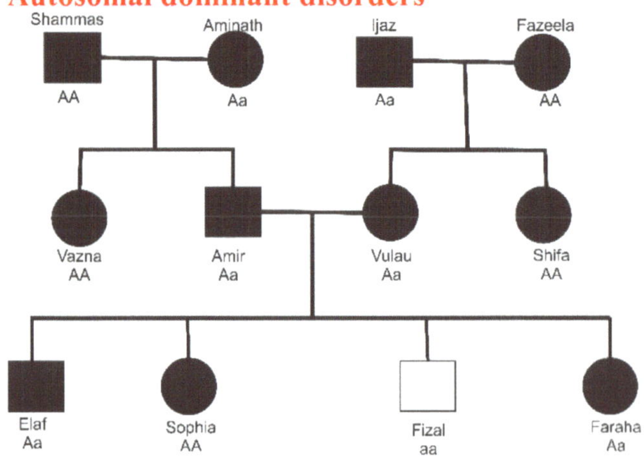

- Fizal (**aa**) has inherited one recessive (**a**) allele from each parent (Amir and Vulau). So, Fizal is normal as there is no dominant allele.
- All other individuals are either homozygous for the dominant allele (A) or heterozygous (Aa). So, they are all affected by the genetic disorder.

Self-assessment (Chapter 19)

Step 1: Visit www.teachingtube.com

Step 2: Sign up for the free course "Edexcel AS and AL Bio assessments"

Step 3: Attempt the chapter Quiz and get instant results.

Step 4: Download the assessment worksheet and marking scheme too.

Chapter Twenty
Cystic Fibrosis

Edexcel IAL from 2018 Specification 2.16: understand how the expression of a gene mutation in people with cystic fibrosis impairs the functioning of the gaseous exchange, digestive and reproductive systems

Cystic fibrosis (CF) is caused by a defect in a gene called the cystic fibrosis transmembrane conductance regulator (CFTR) gene. The gene is active in mucus secreting cells, called Goblet cells. This gene makes a channel protein that secretes chloride ions into the mucus. In people with Cystic Fibrosis, the CFTR protein does not work effectively. This causes thick, sticky mucus which is the main features of CF.
The **normal functioning of the CFTR protein** in regulating the viscosity of mucus by **negative feedback** is illustrated in the figure 20.1 and 20.2.

When the mucus is thin and runny then excess water must be removed from the mucus as described below.	**When the mucus is thick and sticky then water must be added to the mucus as described below.**
1. Na^+ ions are actively pumped across the basal membrane, from the cell into the tissue fluid. 2. Na^+ ions diffuse from the mucus into the cells, through sodium channels in the apical membrane. 3. Cl^- ions diffuse down electrical gradient through the gaps between cells. 4. The movement of ions decreases the water potential of the tissue fluid and water is drawn out of cells at the basal end, by osmosis. 5. Water is drawn out of the mucus by osmosis. So, mucus becomes more viscous (stickier).	1. Cl^- ions are pumped from the tissue fluid in to the cell across the basal membrane. 2, The CFTR channel opens and Cl^- ions diffuse out of the cell, into the mucus. 3. The CFTR channel closes the Na^+ channels, but Na^+ ions diffuse down the electrical gradient into the mucus through the gaps between cells. 4. The movement of ions decreases the water potential of the mucus and water is drawn out of cells in to the mucus, by osmosis. 5. Water is drawn from the tissue fluid by osmosis. So, mucus becomes less viscous (runnier).

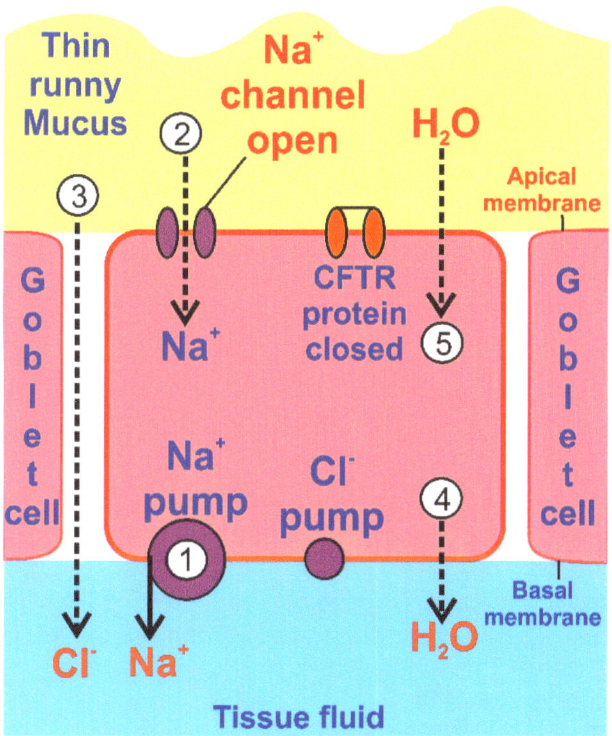

Figure 20.1: Thickening of mucus

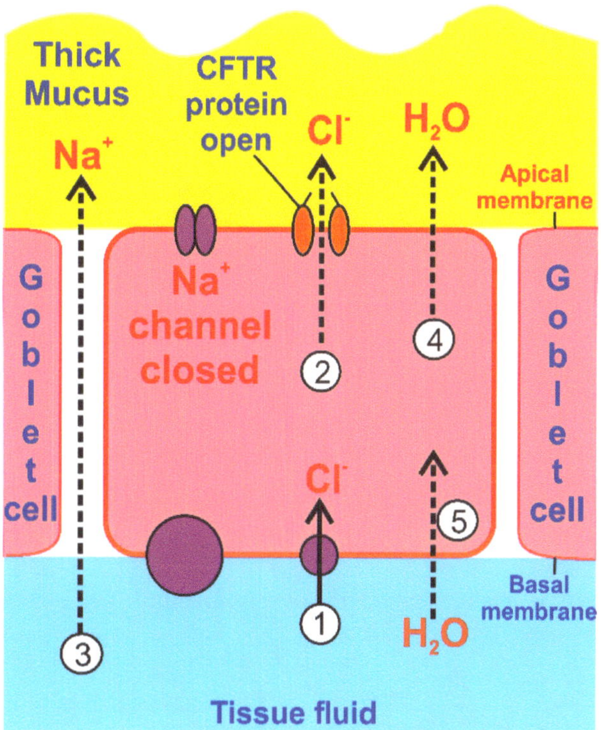

Figure 20.2: Thinning of mucus

Cystic fibrosis is caused by a mutant CFTR gene. The mechanism that is meant to make the mucus runnier does not operate. Positive feedback comes into play and the mucus is perpetually thick and sticky. The mechanism is described in figure 20.3.	
1. CFTR channel is absent or dysfunctional and is not able to secrete chloride ions into the mucus. It also fails to regulate the closing of the sodium channel. 2. **Na^+ ion channel is permanently open** and Na^+ ions continue to enter the cell at the apical end, through the open Na^+ channels. 3. No secretion of Cl^- ions occur. 4. The tissue fluid always has a high ion concentration and low water potential. So, water is continually removed from the cells into the tissue fluid by osmosis. 5. Water is continually removed from mucus by osmosis. **The mucus is always thick and sticky.**	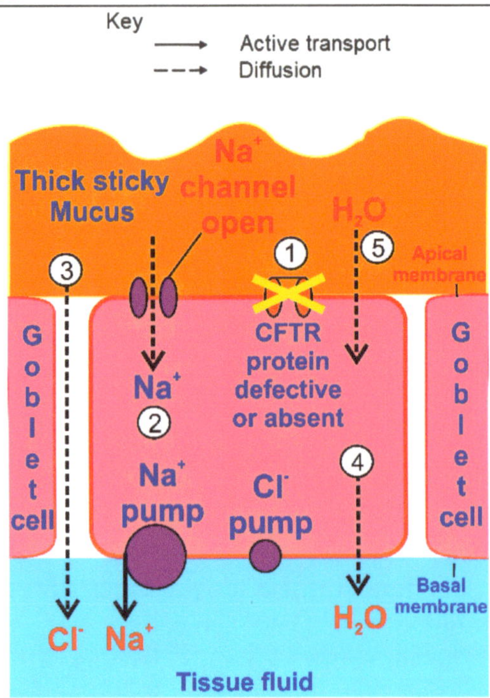 *Figure 20.3: Mutant CFTR results in thick sticky mucus*
Functions of the CFTR protein The CFTR acts as a channel protein to transport negatively charged chloride ions out of the epithelial cells into the mucus. The CFTR protein also regulates the opening and closing of sodium ion channels.	

The effects of cystic fibrosis on different systems are described below.

Digestive system

The thick, sticky mucus can block tubes, or ducts, in the pancreas. As a result, digestive enzymes that are produced by the pancreas cannot reach the small intestine. These enzymes help break down the food. Without them, the intestines cannot absorb fats and proteins fully.

As a result:

- Nutrients leave the body unused, and the individual becomes malnourished.
- Stools become bulky. Some newborns require surgery to remove the bulky stools.
- There may be intestinal gas, a swollen belly, and pain or discomfort.
- Pancreatitis. Pancreatitis is inflammation in the pancreas that causes pain. Cysts appear on the pancreas. That is how the name cystic fibrosis was derived.
- Rectal prolapse. Frequent coughing or problems passing stools may cause rectal tissue from inside to move out of the rectum.
- Liver disease due to inflammation or blocked bile ducts.

Lungs

The lungs are normal at birth but breathing problems can develop at any time afterward. Thick secretions eventually block the small airways, which leads to **inflammation and thickening** of their walls. As larger airways fill with secretions, areas of the lung collapse and contract and the lymph nodes enlarge. All these changes make breathing increasingly difficult and reduce the lungs' ability to transfer oxygen to the blood. Respiratory tract infections occur because of bacterial growth in the bronchial secretions and walls of the airways.

Bronchiectasis is a lung disease in which the bronchial tubes, or large airways in your lungs, become stretched out and flabby over time and form pockets where mucus collects. The mucus provides a breeding ground for bacteria. This leads to repeated lung infections. If not treated, bronchiectasis can lead to serious illness, including respiratory failure.

Reproductive system

People with cystic fibrosis often have impaired reproductive function. Almost all men have a low sperm count (which makes them sterile) because one of the ducts of the testis (the vas deferens) has developed abnormally and blocks the passage of sperm.

In women, cervical secretions are too thick, causing decreased fertility. Otherwise, sexual function is not affected. Women with cystic fibrosis have a higher likelihood of complications during pregnancy (such as developing a lung infection or diabetes), but many women with cystic fibrosis have given birth.

The effects of cystic fibrosis are summarised in figure 20.4.

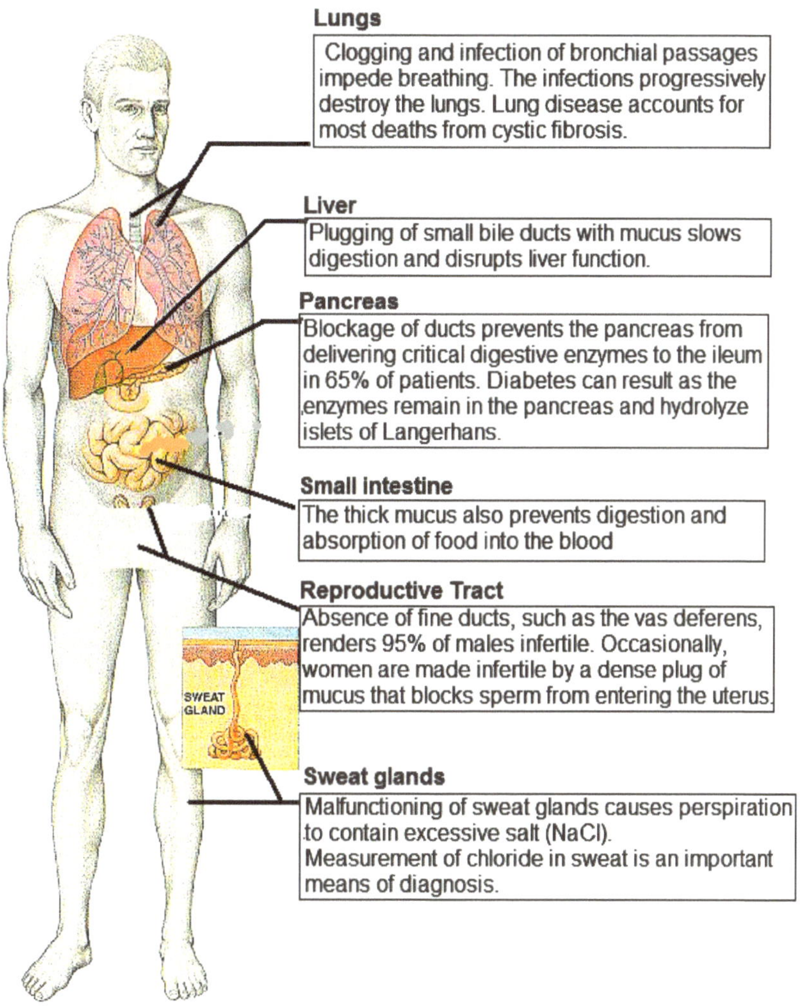

Figure 20.4: Effects of cystic fibrosis on various organ systems

Self-assessment (Chapter 20)

Step 1: Visit www.teachingtube.com
Step 2: Sign up for the free course "Edexcel AS and AL Bio assessments"
Step 3: Attempt the chapter Quiz and get instant results.
Step 4: Download the assessment worksheet and marking scheme too.

Chapter Twenty One

Genetic Screening

Edexcel IAL from 2018 Specification 2.17: (i) understand the uses of genetic screening, including the identification of carriers, pre-implantation genetic diagnosis (PGD) and prenatal testing, including amniocentesis and chorionic villus sampling.
(ii) understand the implications of prenatal genetic screening.

Genetic screening

Genetic screening is the detection of mutant genes by analysis of DNA, using radioactive probes.

Uses of genetic screening

1. Identification of carriers

Since most children with recessive disorders are born to parents with normal phenotypes, genetic screening is used to determine whether a prospective parent is a carrier. This is useful for parents to make informed decisions about having children. In some countries, premarital testing is encouraged to reduce the incidence of heterozygous couples, as they can produce affected offspring with the recessive disorder as shown in figure 21.1.

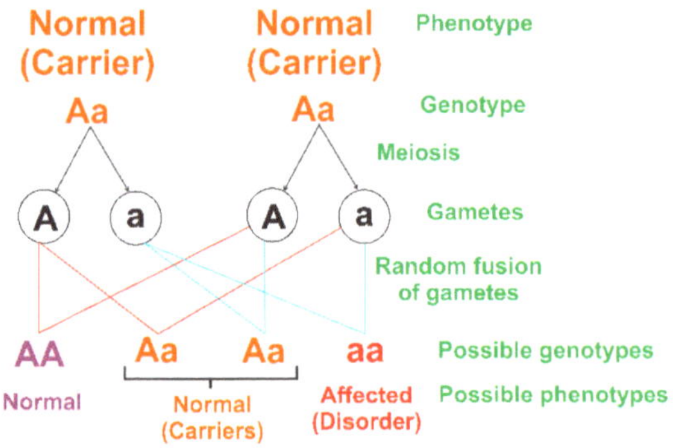

Figure 21.1: Inheritance of recessive disorders involving heterozygous parents

2. Detection of defective embryos

a. Pre-implantation Genetic diagnosis (PGD)

Two carriers of a genetic disorder donate sperm and ova. *In* vitro fertilization is carried out and the embryos are allowed to grow to the eight-cell stage in a Petri dish. A single cell is extracted from each embryo and genetic screening is carried out. Healthy embryos are identified and implanted into the uterus of the mother. This technique allows two carriers to have a healthy normal baby.

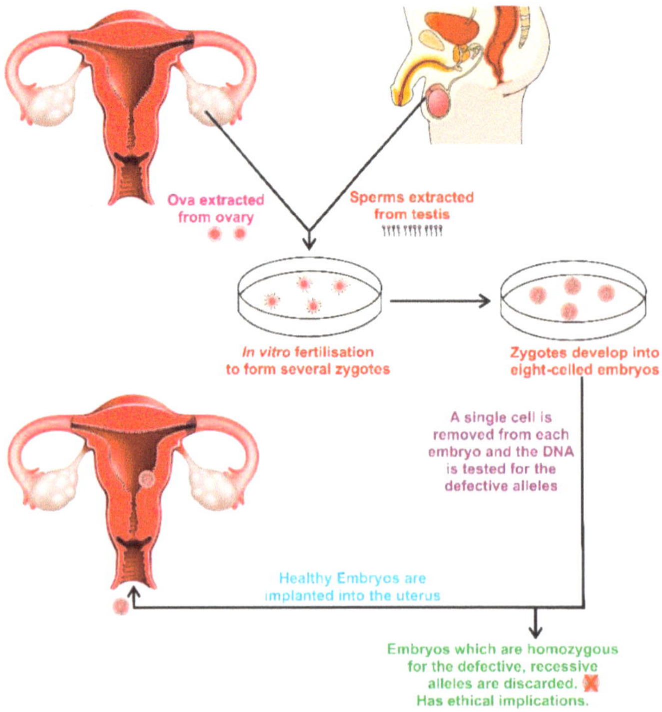

Figure 21.2: Pre-implantation Genetic Diagnosis (PGD)

b. Post-implantation Genetic diagnosis or prenatal testing

Prenatal testing is used to determine whether an embryo has a disease. It helps the parents to make informed decisions about termination of the pregnancy by abortion or gives them enough time to prepare to deal with the arrival of a defective child. It also allows physicians to start treatment at the earliest.

i. Amniocentesis	ii. Chorionic Villus Sampling
A sample of amniotic fluid containing fetal cells is withdrawn through the abdominal wall at about 15 to 16 weeks' gestation. The amniotic cells can be cultured for further tests. Analysis of DNA is carried out to identify specific gene defects. The results from cultured cells are available in about 3 to 4 weeks. The test is reliable and safe with low risk of miscarriage.	Cells are taken from the chorionic villi, through the cervix or through abdominal wall, under ultrasonic guidance. Biochemical and chromosomal tests are carried out on fetal cells. The culturing of cells is not needed, so the results are obtained more quickly than amniocentesis. It is performed at between 9 and 12 weeks gestation. The risk of miscarriage is slightly higher than normal at this time.
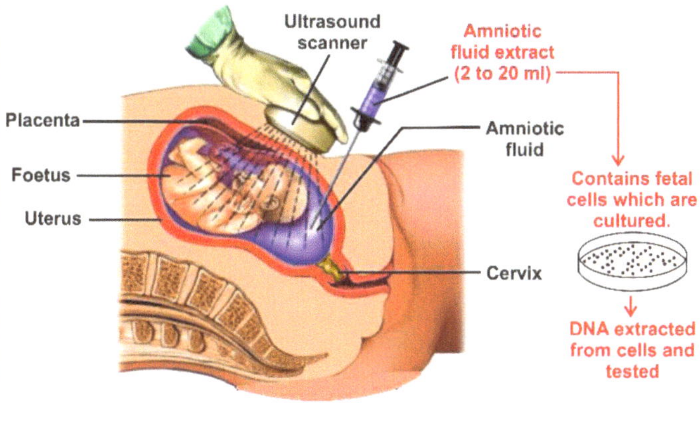 *Figure 21.3: Amniocentesis*	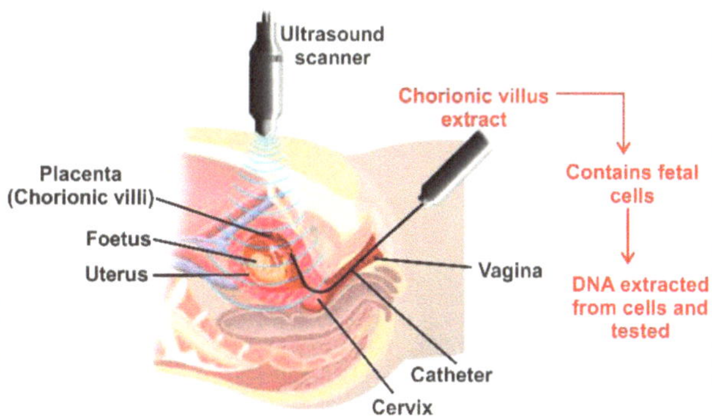 *Figure 21.4: Chorionic Villus Sampling*

Edexcel IAL from 2018 Specification 2.18: be able to identify and discuss the ethical and social issues relating to genetic screening from a range of ethical viewpoints, including religious, moral and social implications

Genetic screening will reveal information that will need to be considered from an ethical viewpoint. It may also have religious, moral and social implications.
Some of the issues against genetic screening are discussed below.

Uncertainty

Genetic testing is not 100% accurate and may provide false positive results. This can result in the abortion of a healthy fetus. On the other hand, if the test provides a false negative result then the parents may not be prepared for a child with the genetic disorder.

Parental disputes

The genetic tests may reveal undesirous facts, like the issues relating to confidentiality of parents.

Abortion and miscarriage

There are ethical and social issues associated with genetic screening of unborn fetuses because this procedure allows parents to **abort** fetuses with a chromosome or gene abnormality. There is also evidence that, in some parts of the world, **abortion** based on **gender** is carried out.

Discrimination – a social issue

If the genetic screening reveals a genetic disorder then, future employers, insurance companies, banks, etc. will discriminate people based on their genetic makeup.

Moral issues

Moral issues are associated with embryo selection, because many unused embryos will be eventually discarded, especially in pre-implantation genetic diagnosis.

Some of the issues in favour of genetic screening are discussed below.

Reduces suffering of victims and families

Genetic screening gives couples **a choice** of whether to allow the birth of a child with a genetic defect.

It is unethical to bring a child with a severe genetic disease into the world if this will result in substantial suffering of the individual, greatly reduce the happiness of parents and family, or drain the financial resources of society.

Allows individuals to make lifestyle changes to deal with the disorder

If genetic screening reveals that a person has a genetic predisposition for cardio vascular diseases (CVD) then the person could make lifestyle changes to reduce the risk of developing CVD. Likewise, if a couple know that they are getting a thalassemic child, they could make psychological and financial preparations to cope with the upbringing of the child.

Self-assessment (Chapter 21)

Step 1: *Visit www.teachingtube.com*

Step 2: *Sign up for the free course "Edexcel AS and AL Bio assessments"*

Step 3: *Attempt the chapter Quiz and get instant results.*

Step 4: *Download the assessment worksheet and marking scheme too.*

www.ingramcontent.com/pod-product-compliance
Lightning Source LLC
Chambersburg PA
CBHW061315170726
48282CB00020B/106
9788191070590